Badiou and Indifferent Being

Also available from Bloomsbury

Badiou and His Interlocutors, Alain Badiou
Being and Event, Alain Badiou
Conditions, Alain Badiou
Mathematics of the Transcendental, Alain Badiou

Alain Badiou: Live Theory, Oliver Feltham
Badiou, Balibar, Rancière, Nick Hewlett
Anti-Badiou, Francois Laruelle
Badiou's 'Being and Event': A Reader's Guide, Christopher Norris

Badiou and Indifferent Being

A Critical Introduction to *Being and Event*

William Watkin

Bloomsbury Academic
An imprint of Bloomsbury Publishing Plc

B L O O M S B U R Y
LONDON • OXFORD • NEW YORK • NEW DELHI • SYDNEY

Bloomsbury Academic
An imprint of Bloomsbury Publishing Plc

50 Bedford Square
London
WC1B 3DP
UK

1385 Broadway
New York
NY 10018
USA

www.bloomsbury.com

BLOOMSBURY and the Diana logo are trademarks of Bloomsbury Publishing Plc

First published 2017

British Library Cataloguing-in-Publication Data
A catalogue record for this book is available from the British Library.

ISBN: HB: 9781350015661
PB: 9781350015678
ePDF: 9781350015654
ePub: 9781350015685

Library of Congress Cataloging-in-Publication Data
A catalog record for this book is available from the Library of Congress.

Typeset by RefineCatch Limited, Bungay, Suffolk
Printed and bound in Great Britain

Contents

Introduction

It is the history of philosophy that determines the transmissible communicability of its concepts, not their truth as such which is, Badiou maintains, universal and, being unexpected, precisely what is not transmitted and non-communicable, as least in its first manifestation. What is intelligible as philosophy is the result of what philosophy perceives itself to be in terms of what it has habitually done and as regards what it wants. In that, since the pre-Socratics, what philosophy has obsessively done is pursue the problem of the One and the many, in the form of the correlation between Being and beings, what is intelligible as philosophy is the procedure of trying to make the reciprocal pairing of One and many function as the ground of its sovereignty over all other disciplines of thought. This procedure is mostly made up of assertions of the Being of the One and observations of the multiplicity of beings, or the many. It is also composed of realizations of the limitations of the scope of the One in relation to the many, and agonies over the imposition of consistency onto the multiple, aleatory and possibly infinite realm of the many, and so on. Over more than two millennia, up to the beginning of the last century, the basic meta-problem of philosophy was stable because that problem was what made philosophy itself communicable as a practice. The components One and many need, for us philosophers, to find their own internal coherence as an articulated, dialectical economy of some or other order. At the same time, the solution to this coherence must articulate the One and the many in a manner which is itself coherent, at least to philosophers. By the beginning of last century, the relation between the One and the many came to be seen as intractable, misguided and poorly posed, so that increasingly both sides of our philosophical tradition abandoned the pursuit of this question as their explicit means of being communicable as 'philosophy'. In a sense, at that moment, philosophy as a single communicable entity sharing both common questions and history along with mutual methodological conditions came to an end. It seemed unlikely that it would ever recover.

With the publication of Alain Badiou's *Being and Event* (1988), the non-communicable nature of philosophy due to its internal rift, its failure to agree on its conditions and its abandonment of its ontological foundations, suddenly appeared less dramatic. Certainly it took years for Badiou's work to make any kind of national, let alone worldwide impact, *Being and Event* was not translated into English until 2005 for instance, but finally a work of unashamed ontology that grappled with the topics of One and many while straddling the intelligible languages of both continental and analytical thought, using a strict formalism, had arrived. Yet on closer inspection Badiou revealed himself to be anything but a demanding saviour of our ailing discipline. His famous axiom, ontology is mathematics, makes it plain that any solutions he may present to the problem of One and many are the result of mathematical rather than philosophical reasoning. Indeed, Badiou is perhaps not overly generous with the portioning out of the sovereign intelligibility of philosophy, denying it any truth of its own other than that of thinking the mutual co-existence of other truth procedures or what he calls conditions. Certainly, mathematics is one of these, but at best what *Being and Event* condones is the application of a communicably sanctioned philosophical language to the discoveries of mathematics, so as to make these innovations intelligible across our diverse community. In this way philosophy is no longer the sovereign but more an ambassador, a role with some pomp but whose glamour comes from proximity to power rather than possession of it. If the axiom 'ontology is mathematics' revitalizes ontology, it does so at the expense of its historic host, philosophy, handing over the problems of the One and many to the professionals to solve, revealing philosophy to have been an assiduous and tireless custodian of the truth of Being, but nothing more. Philosophy has to be happy to pocket its gold watch after a couple of thousand-and-a-bit years of loyal service.

If Badiou is not interested in the philosophy of Being, indeed there is no philosophy of Being there is only the philosophical translation of the mathematics of Being into an intelligible and communicable mode of philosophical discourse, then perhaps we will find a more fruitful field to forage in with the event? The event is Badiou's preferred term for radically singular change, a quality often reserved for the many, although not in Badiou's case. You should note that his masterpiece is called *Being* and *Event* because in interviews he has stressed that the 'and' in this formulation is all important, alerting the reader to the fact that, for all the excitement of ontology, it is the event that concerns him primarily. Further, although Badiou's name has become synonymous with the event, as de Beistegui stresses, Badiou's event has to be read within a wider consideration of the use of the term by Heidegger and Deleuze.[1] Then there is the contribution of Jean-Francois Lyotard to contend with as well.[2]

Even if Badiou could be said to be in possession of the final word on the event, and the evidence is compelling that he is, he has made it clear that this alone does not interest him. He is aware, after all, in conversation with Bruno Bosteels, of the event being a rather common pursuit: 'Let us say that everything in philosophy that is somewhat progressive ... somehow hovers around the notion of the event'.[3] The event is hardly news to continental philosophers, it is, in actuality, the contested ground of modern 'continental' philosophical communicability, but even if Badiou takes ownership of its unstable, indeed perhaps unmappable, territories, he is unlikely to declare himself victor, at least not immediately. In that the event, as we shall see, exceeds the field of ontology, at least the event could still belong to philosophy, but even if that were the case, 'proving' the event is not enough for Badiou. As he goes on to say in the same revealing conversation: 'It is not at all the same thing to say that every situation contains a point of excess, a blank space, a blind spot, or an unpresented point, and to say that this already amounts to the event's effectuation properly speaking'.[4] Rather, he thinks, this would lead to a structuralized presentation of the event which almost contravenes the whole purpose of the event for Badiou, which is to render structures internally inconsistent.

We now have a clear sense of what Badiou is not 'really interested' in. He is not a philosopher of Being but a philosophical custodian of the mathematics of Being. Nor is he a philosopher of the event per se. True, of all philosophers of difference he is the first to promote a formally consistent and thus philosophically communicable theory of singularity and change, and this ought to make him *the* philosopher of difference. Yet he is not *the* philosopher of difference because, as we hope to show, he is *the* philosopher of indifference and the philosophy of difference, without fail, has always found the category of indifference contemptible. It would appear on this reading that Being is only valuable to Badiou because it leads to events, and as events alone have little worth to him, then events themselves must lead to something greater. But what is this rare prize for which Badiou would give up a philosophical king's ransom, ontology and singular difference, to attain? Many would retort the subject but, I would suggest, they would be precipitous to do so.

Interestingly the answer to this question is not to be found within the heavy bounds of *Being and Event* itself. In the years which followed its publication, criticisms of the limited scope of the book, in particular those lodged by Jean-Toussaint Desanti and Peter Hallward,[5] began to take their toll on Badiou's confidence that he had said all he needed to say about the event, and in 2006 he published the follow-up to *Being and Event*, *Logics of Worlds*. It is accurate to say that *Logics of Worlds* has not had anything like the impact of *Being and Event*, and even after its publication many studies of the event and related concepts such as the subject, politics and truth have appeared which mostly regard the second volume as an interesting development, but nothing more. Indeed, if any other Badiou text is having a bit of a moment, it is the earlier, more avowedly Marxist, *Theory of the Subject*,[6] which perhaps gives us some indication as to the current timbre of Badiou's influence. The undervaluing of *Logics of Worlds* is perhaps the only glaring error within the brilliant Badiou community in that, truth be told, *Logics* changes everything about the event as a theory of change and without a detailed engagement with its philosophical architecture, you really do not 'get' what Badiou is about at all. Don't take my word for this, Badiou says so himself in the same interview with Bosteels back in 2005, a year before the publication of *Logics of Worlds*.

In a section called 'Replying to Critics' Bosteels is trying to come to terms with the reception of *Being and Event* amongst 'radicals' and Badiou, laughing, makes the point again about the importance of the term 'and'. Bosteels takes this 'and' to be disjunctive, to be read almost as Being vs. Event. Badiou's world of being is, on the surface, overwhelming in its tedium, intended in part to make the event an itch you cannot help but want to scratch. Yet Badiou disagrees with Bosteels' reading. 'It is not the opposition between the event and the situation that interests me,' he explains. '[T]he principle contribution of my work does not consist in opposing the situation to the event ... The principle contribution consists in posing the following question: what can be deduced, or inferred, from there from the point of view of the situation itself.'[7] He goes further along this trajectory when he questions how some have called his commitment to the event 'dogmatic' when in every case we 'are in the register of consequences' or, the event as such does not matter only the consequences the event has on an already existent situation.

Badiou ends the discussion of the reception of his work up to that point with the following, I would say still-current, overview of the real importance of his life's work: 'Really, in the end, I have only one question: what is the new in a situation? My unique philosophical question, I would say, is the following: can we think that there is something new in the situation, not outside the situation nor the new somewhere else, but can we really think through novelty and treat it in the situation?'[8] To do this, he notes, we have to first come to terms with what a situation is or 'what is not new, and after that we have to think the new'.

We have our answer therefore. What matters to Badiou is the situation such as it is in terms of the new that occurs within it. And we also have the pathway of our own study laid out before us. Before we can think the event, we need to understand the situation, and only after we understand the situation can we understand the event, not in terms of its intrinsic nature but by virtue of the means by which it changes, radically and irreversibly, the situation within which it occurs. Yet we have to be clear and careful here. For example, although Badiou himself uses the term 'situation' to refer to everything that 'is', the realm of Being as we have come to understand it since the Greeks, within a year he will instead call a situation a 'world' and what people understood as a situation up to that point will no longer hold. Naturally it follows from this that if he changes the meaning of 'situation' and an event's significance is defined as its consequences within a situation, then his theory of the event will have dramatically altered and this is certainly the case for the simple reason that world is not just another term for situation, but a complete reconceptualization of the situation from the perspective of category theory and intuitionist logic, two elements that have absolutely no purchase on the concept of situation such as it is presented in *Being and Event*.

Secondly, we need to be very attentive to the disjunctive conjunction 'and'. According to Badiou, there are no events in the realm of ontology. So Being does not 'need' events. This would suggest that events are opposed to Being in some way. Yet Badiou says very transparently to Bosteels that events alone are of no interest to him unless they change a situation from the inside. So now 'and' is conjunctive: being + event = radical change to a situation. Yet a situation is not the same as saying Being in actual fact. Different ways of reading 'and', as disjunction, conjunction, compossible, causal, co-relational, nonrelational and so on will determine our ability to grasp if an event, due to its consequences, by virtue of the fidelity of subjects in pursuit, retroactively, of an event's truth due to the changes it has brought about in the continuum of what already is, not only exists, but is worthwhile dedicating your life to its demands, as Badiou says he has committed his life to the consequences of the event called Paris '68.[9]

There is, I suppose I am arguing, a 'new' Badiou whose likeness has not yet been captured. I am busily sharpening my charcoal in order to make the best job of it I can. This is part one of our two-part thesis which goes as follows. Thesis 1: You cannot understand *Being and Event* unless you come to terms with the problem of their 'relation', in particular the importance of the event in terms of its situational consequences, which themselves are only possible to conceive of after the systematic reconsideration of the situation as a world due to *Logics of Worlds*. The second part goes as follows. Thesis 2: In that what matters is not that the event is 'different', and because of the fact that Badiou openly rejects all considerations of singular difference posed by the philosophers of difference, then the 'difference' of the event is meaningful only in relation to Being and situations. In that Being, multiples and, up to a point, situations/worlds are, due to set theory, *indifferent*, the difference of the event, the difference it makes due to consequences, has to be seen against the backdrop of indifference not, as has been the tendency in our community since the sixties, against a backdrop of identity. This, at least, is our reading of Badiou's comment that first you start with 'what is not new, and after that we have to think the new'. We won't be able to think the new from the pages of *Being and Event* alone, but before we tackle *Logics of Worlds* we need to concede that the 'not new', Being as such and its worldly situations, is defined by its being indifferent. In short, if the event is what interests you, then the event needs to be read after *Logics of Worlds* due to a reading of Being as indifferent, and this is what our two-volume study proposes. First understand the not-new as indifferent, and then understand the new as radical change within the absolute communicability of worlds.

The consistency of inconsistency

Badiou's ontology proffers a deceptively simply proposition. He allows things to be inconsistent through the application of formal procedures which themselves are profoundly stable, reproducible, and indubitable. In doing so, he is able to accept inconsistencies, which ontology in general has baulked at, because access to them is provided by communicably unquestionable formal consistencies, such as recursive deduction, classical logic, axiomatic reasoning and set theory. What this has permitted him to pursue, in terms of Being, beings and, to a large degree, events, is a position where their intractable incompletions or inconsistencies are fine, as long as we construct them in languages and logics which are themselves very, very, albeit not absolutely, consistent.[10] Take Being for example, the consistency par excellence of our tradition. Rather than articulate the One of Being and the many of beings in terms of stable identities and their differences, instead Badiou proposes that we prove the procedural 'consistency' of a permanently inconsistent One due to the stable recurrent counting of some ones, such that we can even speak of an infinite number of these ones, usually the cue for a philosophy of chaotic multiplicity, in such a way as we can always be sure of their internal coherence, even though we cannot count their externalized completion. Badiou achieves this by downgrading the big 'O' One to the much less beguiling count-as-one, a unified collection of ones within a specific situation guaranteed by a formally stable set of procedures, a count, so that wherever you are within a particular world, and you are always somewhere in his immanentist philosophy, the components of your location combine into an absolutely stable, yet open-ended, state or set. Or, to put it another way, every situation or world is a stable one, even though no situation is a closed-off and consistent transcendent One.

In Badiou's work, the One suffers the slight ignominy of being down-graded to the count-as-one; does this mean by the same gesture that the many, or difference, experiences an up-grade into a truly singular difference, what Badiou calls the event? Surely that is, like Deleuze, how he does it? Maybe, but before we get there like Being or the One, difference, or the many, first has to be reconstituted. In that One and many are reciprocal, changes to one party inevitably recoil onto the other. The many, for example, is no longer difference per se but the consistency of the count-as-one; Being is simply that which allows an entity to simultaneously be counted as a being and as part of a being. If this is the case then difference as we have taken it to be for the last fifty years, a means by which multiplicity can undermine the stability of Being, has lost all of its bite. Instead, difference is the saviour of Being in that counting as one is simply the ability to set a limit to the multiplicities of our worlds, without having to argue that they are totalizable, transcendentally sanctioned or internally complete. Put simply, difference, taken as the many, *is* Being, the countable stability of an incomplete ever-changing situation, or at least one way of saying Being. The collapsing of Being into the many through the formal countability of even an infinity of beings, appears to be the death knell of difference, properly conceived, but in many ways this is just strategic on Badiou's part. He may not think much of difference as multiplicity, but one thing he does see as valuable is the way in which we are forced to concede that difference, true difference, has to be something much more than alterations within otherwise stable worlds.

Show that difference exists, real difference in particular, called the event, because it is inconsistent with Being as regards what counts as one in a manner which is, however, retroactively, in terms of its consequences, profoundly consistent: this is the second application of the consistent inconsistency process Badiou innovates. That which we took to be consistent, Being as such, is now proven to be inconsistent, there is no total, transcendental, complete One, by using formal, deductive tools which are themselves almost unbearably stable. While that which we have celebrated as inconsistent, our modern philosophy of difference, is shown to be actually not inconsistent at all. Instead, a truly singular difference, an event, can occur in a manner that makes it radically inconsistent with the worldly situations which exist, but at the same time this event can be shown to be profoundly consistent in a manner which does not negate its radical singularity but indeed extends and strengthens it. A consistently inconsistent Being allows for an inconsistent event whose consequences in the world are themselves formally and procedurally consistent. This play between the consistent and the inconsistent is, for me, the real dialectic at the heart of Badiou's thought: through consistent modes of thinking, that are communicably intelligible and transmissible among us, remove the ontological and wider philosophical taboo, in our discipline, of open-endedness.[11] When Badiou says Being inconsists, he means that it can 'exist' as inconsistent absolutely, permanently, everywhere because the means of showing this are indubitable, or as indubitable as any other scientific assertion. And when Badiou says that an event 'exists' he means it is singular to a consistent situation because its inconsistency within that world can be traced through the consistent procedure of assessing its consequences. In each instance, in different yet related ways, Badiou's philosophical method is that of accepting inconsistencies through procedures which themselves are powerfully consistent.

Subtractive being

There are two central tenets of *Being and Event* that unlock the rest of the text. These are that being is-not or is 'subtractive', and ontology is mathematics. What the two statements say when articulated is that the age-old problems of ontology are solvable if you say that being is-not rather than it is, and the mode of formal reasoning that allows such an impossible statement to be made – how can you say being is-not when ontology is the science of Being as what is – is mathematics, in particular set theory. If we take ontology to be the science of everything that is, after Badiou what we are really speaking of is ontology as the consistency of everything that is, due to the mathematics of what is-not. Ontology, therefore, is not so much the science of what is, as the mathematics of what is-not.

Set theory provides the proof for subtractive being but before that we need to be clear what we mean by the impossible statement being is-not. Being is taken as the unassailable consistency of a field or situation of existence. Being is traditionally read as the universal whole, against which local beings define their existence as being examples or elements of that whole. If we are all different in some way as human beings, we are all the same in saying that we are all different kinds of human being. We might even say that to be a human being is to possess the self-consciousness of this relative difference. For the record, this simple model has never worked.

Start again then and ask what Being represents for us. Being determines the possibility of the consistency of beings taken together as elements of a commonality, we will call this a situation. Mathematically, what we are speaking of is a consistency of multiples as being all part of the same set. Being then is a guarantee of the consistency of multiple beings all spoken of in the same manner: *human* beings, *animal* beings, *mineral* beings and so on. Thought of in this way, Being as such does not have to *be any thing*. All it has to do is confer consistency to a situation of beings. Being, therefore, is not a thing but a process, specifically a process of counting. Realizing this, Badiou says that if you accept that being is the precondition to the stability of a situation, being does not have to be included in that situation. This in the past was the main sticking point as it inevitably led to self-predication and Russell's paradox.

There are two ways to proceed from this. The more complex is to explain how Badiou realizes, due to the axioms of mathematical set theory, that if you define being as not present to any situation such that the situation can be consistently countable, this is the same as saying being is the void subtracted from any set of multiples such that the set in question is consistent. Being is subtractive in this way because it is removed from the count so that the count will always be consistently stable. This is the 'proof' that being is-not.[12] The easier way to say this, I think, at this early stage is that given any situation as a collection of elements all said to belong to that situation, you can prove that immanent situations are completely consistent, without imposing a transcendental whole onto that situation. After all, Being as such was only seen to be needed to make sure that multiplicity, which is empirically and intuitively a given, does not proliferate into chaos or bad infinity. Being is just a means of talking about beings in a consistent, transmissible and communicable fashion. Indeed, Being was always subtracted from a situation, that was its problem. Being was always the abstract definition of its beings so that that Being as such could never be one of those beings. This kind of Being, by the way, could never be proven and so could only ever be spoken of in terms of being God-like or in-withdrawal.

If you say being is-not, you are saying an infinite number of situations exist, all of which are actually infinite, there is no upper limit to how many elements can join that situation, in such a way as they are totally consistent, not because Being sits above them all as a transcendental title for a situation it defines yet which it can never belong to, but because Being is simply an element subtracted when you count those elements together, such that you can say they are consistently parts of the same situation. When you say being is-not, you say consistent and limited situations exist, which are not, however, completed wholes. Being is-not is the same as saying there are consistencies which can be taken as one single thing, which are however never complete wholes. Badiou's message is simply that you can be a consistent 'one' without ever being a unified, complete, transcendentally whole One.[13]

Nonrelationality

Both Being and event are 'nonrelational' and this aspect of their make-up is the single most crucial point for everything we go on to consider.[14] Having said Being and event are both nonrelational, they are not nonrelational in the same way with the same effects. To speak of the nonrelationality of something, the terms of relation themselves have to be described. One of the most important stipulations across the whole of the Being and Event project is that there is a fundamental difference between the realm of Being determined ontologically due to the axioms of set theory, and that of beings determined logically due to the axioms of category theory. The nonrelationality of Being is determined by what relation means ontologically, while the nonrelationality of beings is subject to the definition of relationality logically. All of which is made a tad more complicated by the long failure in philosophy, perhaps exemplified by Kant's object X, to relate these two realms together so that the nonrelationality Being and event seem to share in common, is in fact composed across two fields themselves taken to be nonrelational with each other. That is one of the larger questions of the two books taken together, the means of their articulation due to what Badiou calls the onto-logical. We will return to it at a much later date.

Speaking of Being first, it is nonrelational because it is-not. We will also call this the in-difference of being or the inability to define it as different because difference is a relational comparison and being is nonrelational. Relation, ontologically, is defined in terms of the elements that can be counted as belonging to any situation. As being is-not, it is not counted and therefore no relation to it can be established. In terms of the count, what relation means is that the objects in the count have no other property, no quality, except where they occur in that count. The being of 2 is determined by it being next after 1 and just before 3. Ontological relation means specifically ranking and is a quantitative determination only. In that being is-not it has no rank and so is non-relational in terms of the local sequential nature of any ranking order.

Events also have no 'rank' ontologically but this is irrelevant because, as we have seen, they are banned from ontology even if they are proven to be possible by ontological axioms. Events present themselves to the logical field of worlds and situations in such a way as they are nonrelational therein. The relationality of a world is very different to that of the relationality of being. Beings do not exist in worlds due to their rank, but due to their comparative relational measure as regards all other beings in that world. If relation in ontology is a single, graded line, in terms of worlds it is a complex of triangular comparisons of two objects relating to each other in terms of how each relates to a single larger object they have in common. In that these relations are differential, qualitative determinations, object *x* is smaller than object *y* relative to object *z*, for something to be nonrelational, an event, they must be without quality. Again, we will call this indifference, in this case the indifference of quality neutrality such that a comparative difference between the element in question and another element cannot be established. There is a clear difference between the in-difference of being and the indifference of event relative to their nonrelationality. Being is in-different because it cannot be presented to relation, to the count of rank. The event is indifferent because being present to a world, it cannot however be differentiated from the other elements of that world due to this difference being a qualitative relation and events possessing no particular quality.

Indifference

The title of the first volume of our study contains the phrase 'Indifferent Being'. In considering the nature of being, then, we will also argue that it is indifferent in essence, a stipulation that is essential and also currently missing from Badiou scholarship. The history of indifference is as long as that of philosophy itself, primarily because indifference has always menaced the foundations of ontology as regards identity and difference. For our purposes we begin with Hegel and his theory of indifferent difference or pure difference as such. In the first section of the *Phenomenology*, Hegel repeatedly considers *pure difference as such* or abstract difference irrespective of any qualities held by the two identities being differentiated.[15] This is best represented by the formal notation A ≠ B. What A and B are in actuality, what qualities they hold, is indifferent, does not matter, they are mere abstract notation. In indifferent difference, all that matters is that there is a difference, A ≠ B, as the basis of the fact that there is an identity, A = A because A ≠ B. This is the founding formula of Hegelian dialectics and all the philosophies of difference to come because it develops a pre-identity form of difference: abstract difference irrespective of qualities.[16]

Yet as is ever the case in the philosophy of difference, of which Hegel is the father, indifference eventually must be negated for difference to flourish. Hegel's construction of the dialectic in terms of difference is founded on the need to negate indifference. In the opening pages of the *Phenomenology*, and consistently through his *Logic*, Hegel considers objects that are singular, non-divisible, to such a degree as we cannot speak of them as kinds of objects in such a way as their existence is communicable among us. At the same time, he worries about wholes which are consistent, only to such a degree that they are consistent truths of no objects per se. His dialectic is an ingenious but flawed attempt to rescue philosophy from a world of objects that are so different we cannot determine the nature of their differences because we cannot compare them, and a whole which is the truly consistent Being of all beings because, in reality, it is the Being of no actually specific beings. The combination of the dialectic plus God saves Hegel from the terror of the indifference of pure determination, every object is a monadic object unto itself, and of total nondifferentiation, the whole is simply the whole of itself, with no elements within it one can speak of. The problem being God clearly does not exist, while indifference resolutely does.

The issue is more clearly articulated in Deleuze's *Difference and Repetition* which is the prime inheritor of the Hegelian tradition and one of the central texts on indifference in the canon.[17] We need, Deleuze says, a theory of difference such that it precedes identity.[18] One option is that differences are taken as a field of pure difference as such, each element within the field being singular but in a quality neutral fashion such that the field is totally abstract. However, we want differences to make a difference, we wish singularity to be a unique combination of qualities. What we do not want is a field of abstract difference where every element is different in the same way. This first form of indeterminate difference is the basis of what we will call indifferent difference due to quality neutrality. After Hegel we can say that $A \neq B$ without knowing what A and B determine as regards their differential predicates, but left to its own devices this results in the pure abstraction of a Being which has no named or specific objects.[19] We could then travel in the other direction and suggest that once determined, all elements are truly different from each other in that each is totally closed and determinate sharing nothing in common with any other element. Again this defeats difference, which is based on a series of relational differences. These closed, singular monads are so different that they do not participate in relational difference and again Hegel speaks of these as totally determinate objects in the world in possession of no unifying commonalities. Effectively every day, every scene would be totally unique, each tree seen for the first time and so on. Not only does this go against the purpose of ontology, atemporal consistency, it appears counterfactual as well, we don't operate like that as beings. This form of radically non-relational difference we will call in-difference. An element is in-different, according to Badiou, when it does not enter into any identity-difference correlation because it is entirely nonrelational to that process.[20]

The two kinds of indifference in play here, indifference and in-difference, are regularly used by Badiou throughout *Being and Event*, a fact rarely commented on. Badiou's genius is, rather than run from indifference, to embrace it. What we will see in due course is that in defining beings or multiples as indifferently indeterminate due to the void of being as in-differently determinate in its isolation from all that exists, Badiou is able to combine indifference as quality neutrality with in-difference as absolute nonrelationality to create an unassailable ontology. Or rather, he is not able to combine content neutrality with nonrelationality, instead he realizes that modern set theory had already combined the two in such a way as first, a formally viable and constructible ontology could be composed, second this could not preclude the possibility of a real event, and third, this event could be 'presented' in terms of its effects in the stable situations we can compose, due to the stable ontology of a being that is-not. This, then, is what the two volumes of *Being and Event* set out to do, an impossible task without a complex and sustained engagement with indifference.

Set theory

Badiou's entire philosophy since *Being and Event* relies on set theory, a branch of mathematics based around the study of collections of 'objects'.[21] Set theory was developed in Germany by Georg Cantor, with significant interventions by Richard Dedekind, towards the end of the nineteenth century. Cantor's theory of sets, based on the assertion of an actual infinity, was subsequently formalized into an axiomatic system by Zermelo and Fraenkel.[22] It is now a totally orthodox part of day-to-day mathematical operations. Set theory is composed of one primitive, multiples that belong, and a relatively small number of axioms, which is the basis for its widespread success. Badiou's favoured version is the eight axioms of Zermelo-Fraenkel set theory plus the axiom of choice, first presented in 1921 (ZF+C).[23] In contrast, Badiou is openly critical of Kurt Gödel's constructible theory of sets, first proposed in 1938, with major additional contributions by John von Neumann, that makes do with seven axioms dispensing with the axioms of foundation and choice. This is not surprising as foundation and choice are perhaps the most important axioms for Badiou's theory of the event. Another key figure in the history of set theory for Badiou is Paul Cohen, whose work in the 1960s on choice in particular, and the process of forcing, led to the development, along with W. Easton, of the theories of the conceptless choice, the generic set and the process called forcing.

One of the most important discoveries of set theory, that which leads Badiou to speak of the Cantor event,[24] is its proof that actual infinities exist. Badiou realizes that set theory, its being concerned with how parts, multiples, combine to form stable wholes, sets, is a potential formalization of the basic proposition of ontology. In particular, set theory's proof of actual infinity and its designation of a smallest element, the void set, meant that its immanentist mode of ordering sets in terms of other sets, themselves ordered in terms of other sets in an apparently impossible, proliferating circularity, simply worked; it was possible. The consistency of an immanentist ontology, always attractive but never achieved, is provided if there is a provable halting point and infinite upper limit which can be said to be actual without its being countable as a closed-off whole.

Set theory axioms present solutions to longstanding philosophical problems but they do at least four other things for Badiou. First, set theory has a key role to play in the philosophy of mathematics, with such questions as do sets/numbers exist?[25] The philosophical basis of sets, then, presents several important conceptual opportunities for philosophy as a whole. Second, the retroactive axiomatic method typical of set theory, although not its exclusive preserve, becomes a means not just of proving the event, but a model of how events can be said to be possible. What mathematics does is propose and then retroactively prove evental truths. Third, set theory allows one to found a very consistent theory of existence due to worlds, or category theory, the formal basis of *Logics of Worlds*. Category theory is not the same as set theory but set theory is a foundational element of category theory. Finally, fourth, due to the inter-relation of set and category theory, eventually Badiou is able to show that actual events exist in worlds through category theory, because of certain tenets of set theory. These last two points will detain us in the second volume; for now, the reader can park category theory and focus entirely on set theory.

Set theory is a mode of mathematical deduction based on the relation of multiples to other multiples. It is concerned with two kinds of counting. The first is how to count a multiple as one, as a single unit, as the thing it is. The second is how to count a multiple as part of a larger one, so that the thing it is in the singular, can be taken to be a component part of the thing something else is, a larger thing, such that said larger, compound thing can also be counted as one.[26] The single count we can call a multiple, the larger count we can call a set, although in the end all sets are composed of multiples of multiples and all multiples are multiples of multiples, so all sets are multiples and all multiples are sets. If you take this to represent pure immanence as we have described it above, what set theory does is provide a formal means by which the immanence of multiples of multiples can be shown to be a procedurally consistent inconsistency, which we are taking as basically consistent.

While Badiou will use Zermelo-Fraenkel set theory plus the axiom of choice, which is composed of nine axioms, in reality you only need three parts of set theory. The first is the axiom of the void set, which is the smallest unit that can ever be counted as one. Second, you need not so much the axiom of the multiple, there isn't one, but to think of the various axioms of the multiple as proving that a multiple is simply that which counts-as-one in a situation such that sets of multiples exist and behave in a consistent manner. For multiples of multiples to exist consistently you must have a smallest multiple, proven by the void set plus the axiom of foundation, and you must deal, third, with the problem of infinity. Clearly the phrase 'multiples of multiples' is inconsistent as it proliferates out to an uncountable level with no upper limit, bad infinity as Hegel called it. If infinity in general is immeasurable, bad infinity is immeasurable just because it is too large a number to count. In contrast, good infinity is a transcendentally stable, immeasurable upper seal on the nature of all things, the Being of all beings, a role traditionally occupied by God. Small problem, God doesn't exist. Solution, what if bad infinity were not so bad after all? What if the uncountable nature of infinity, that it has no single upper limit, were not bad but in fact the best of all possible worlds?

After Cantor, the axiom of actual infinity states that actual infinities exist. You cannot count them, but you can count based on them and you can use the basic rules of this count to prove their stability. Which is another way of saying that just as being is simply a name for that which facilitates the stability of any count, infinity is the name for the fact that every count is provably stable, irrespective of how many elements you may add. The truth is that all situations are uncountable as an overall, closed and consistent One, without the imposition of an inexistent God, but this does not mean that they are not consistent per se in other, formal, transmissible, procedural and secular ways. If, instead of an inexistent God you spoke of inexistent Being as void, you can absolutely prove the overall stability of any situation whatsoever by simply saying it is an actual infinity.[27] In that the phrase, multiple of multiples, is the same as saying, up to infinity, if actual infinities are proven to exist, then all situations can be said to be stable without saying they are wholes, if you admit that all situations or worlds are actual infinities.

These three qualities, the void, multiples of multiples and actual infinities, define the stability of every situation. Every situation can be proven to be composed of a smallest unit, in relation to another unit, in a non-closed fashion which is, however, provably stable. This is what the combination of void, multiple and infinity produces. To sum up, in that all of these deductions depend on the role of the void, which cannot be counted but whose operational presence at the count, due to the axiom of the void set, we can always assume, we can say that everything 'is' in an infinite yet stable fashion, because the void or zero is-not, is present as subtracted from the count. We will go on to show this in extensive detail, but for now we will take it as read.

Retroactive axiomatic reasoning

A particular quality of set theory is its reliance on retroactive axiomatic reasoning, a mode of formal reasoning that, I would now expect, will become a central part of the communicability of contemporary philosophy. In mathematics, a proposition is taken to be impossible due to the current state of mathematical axioms. The mathematician, then, as a thought experiment, asks what would it mean if this proposition were not impossible? What axioms would there need to be to allow this impossible proposition to be possible? Once you have developed these axioms then you have to do two things. First, make sure that the axioms, which are new, do not contravene the conditions of mathematics as a whole, in particular adherence to regressive recursive deduction and its already sanctioned axioms. Second, if you are proposing a new axiom you must ask: What is its use for the community? Does a new axiom solve issues that mathematicians want to solve, and can it do so by replacing other axioms by doing the same job they do, plus bringing to the table these new mathematical benefits? This is what is meant by retroactive axiomatic reasoning. Propose a new axiom, test its methodological acceptability (recursive deduction) and show that if said axiom were accepted to be the case by other mathematicians, said mathematicians would be able to do things they couldn't do before, things they have always wanted to do, or things they didn't think they ever could do, without negating the main formal basis of mathematics such as it was taken to be up to this point. What was initially impossible is then, retroactively, proven to always have been possible due to the development of new axioms that are in accord with the overall values and conditions of the community of mathematicians.[28]

Retroactive axiomatic reasoning is a workaday part of all mathematical and much scientific reasoning, but it has particular resonance for set theory. Cantor's set theory was able to prove that actual infinities exist. Actual infinity and set theory as a whole revitalized mathematics, opening it up to an immense world of new possibilities for mathematicians that would take decades to exhaust. Cantor basically said that if he could prove the axiom of actual infinity through the development of new axioms that allowed mathematicians to achieve new and desirable results without abandoning results they already held dear and without contravening the conditioning formal laws of maths, then actual infinities exist.

What Cantor did was make his axiom of infinity communicable among his community.[29] To do this his axiom spoke to the community in two intelligible ways. First, it addressed them as a community with a common, formal foundation: recursive deduction. Second, it hailed them as a community with common aims, desires and daily activities. If Cantor had presented axioms which were recursively deductive but which did nothing for the community, these axioms would not have been taken up and would have failed to become evental. Instead, if he had presented axioms which did not conform to the basic conditions of mathematical activity, these axioms would have been rejected as not being mathematical. Perhaps they may have been taken up by another discipline, logic or philosophy perhaps. Set theory went on not just to be acceptable to mathematicians, it became the dominant mode of mathematical activity for decades so that from being accepted as a communicably new thing against the background of already existent communicable conditions, it came to be that very background, a fact that Badiou finds inspirational as a model for how an event, the Cantor event, can change a world, mathematics, from the inside, using the sanctioned formal conditions of mathematics in such a way as it answered the communal needs of mathematicians.

If we now transpose this method to ontology, we can see how effective it is. Badiou says that since the dawn of philosophy ontology has been defined as the reciprocal relation of One and many but the only way this relation can be retained is if we say that being is-not. This is an impossible statement among us as philosophers because ontology is the science of what is. Take the impossible proposition, being is-not, to be true all the same. What do you gain? You purchase the possibility of ontology as the reciprocal relation of One and many which is what you wanted. How do you achieve this? By proposing two new axioms of being. First, that the One is actually an operational count-as-one. Second, that the many is actually an indifferent multiple of multiples. What does this allow you to do? You can retain the communicability of philosophy concerned with ontology, you can heal the schism in philosophy opened by twentieth century thought and, eventually, you can retain a Theory of the Subject due to the event. In so doing do you contravene the basic conditions of philosophical thought? This is perhaps the sticking point.

To attain the axiom being is-not you need to state that ontology is mathematics, a position which initially does not seem communicable among us. Both Heidegger and Frege in their different ways and for different communities, want to reject mathematics as the basis for philosophical thought such that this rejection has become a defining feature of the communicability of what we have taken philosophy to be for about a century. But is the assumed non-transmissibility between philosophy and mathematics actually true? Badiou is ever-ready to point out that philosophy was always mathematics and, due to analytical philosophy, remains substantially formal.[30] If we have forgotten this it is mainly the fault of Hegel, Nietzsche and Heidegger in turning continental philosophy away from its essential, deductive conditions, and the anti-realist strand of analytical, constructible thinking. Coming back to our thought experiment then we can now say yes, with some revisions, we can retain the axioms being is-not, the one is count-as-one and the many is an indifferent multiple of multiples, in that they allow us to do something we could not do before, and yet they do not negate the conditions for the legitimacy of our saying that what we do means we are all philosophers if we accept that historically all philosophers are geometers, as Plato famously set in stone.

One final point needs to be made here. In that we take the one as operational, and the multiple as indifferent, we are actually using axiomatic reasoning to define the very axioms of ontology. In axiomatic reasoning it is the operational sustainability of an axiom that defines its truth: Can this axiom be used repeatedly by many different mathematicians with the same results and as the basis of entirely new directions in their reasoning? Here we ask: Can the count-as-one be a truthful statement about being not due to what it says about what being is, but due to what it allows the theory of being to actually do? Further, in axiomatic reasoning the specificity of the axiom is secondary to the indifference of its abstract formality. For a mathematician an axiom is not a thing in the world or even a truth, but merely a caption for a set of abstract modes of reasoning which, if necessary, can all be reduced to the indifferent neutrality of numbers and operations due to recursive deduction.[31] What Badiou calls a model in his earliest work, *The Concept of Model*.[32] So not only are the new axioms of being made possible by retroactive axiomatic reasoning, but their very nature is made possible by axiomatic thinking as a general process. Being is-not because it is operationally counted as present in an indifferent, abstract process of reasoning. In short, the axioms of set theory provide us with the proof of ontology while the general axiomatic method, not specific to set theory, presents us with the model of how the basic axioms of being can become communicable amongst us as a replacement for the idea of being as based on the location of objects in relation to their class.[33]

Transmissibility, intelligibility and communicability

Central to the success of retroactive axiomatic reasoning is that whatever it proposes will be able to demonstrate unquestionable, mathematical transmissibility. Such transmissibility determines what we are calling discursive 'communicability'.[34] The primitive maxim of communicability, which we take from Foucault's theory of intelligibility and Kant's subjective universality of aesthetic judgement, is the meaning of any statement is not what is said but that it can be said.[35] The content of the statement is meaningless alone, but its participation in a consistent world where this new statement can immediately be understood is not only meaningful, it is the truth of all meaning. As communicability is the topic of our second volume, *Communicable Worlds*, we will keep our powder dry to some degree and only speak of communicability and intelligibility when it pertains specifically to the issue of being, in particular as regards the nature of retroactive axiomatic reasoning and the transmissibility of set theory axioms across the community of mathematicians.[36]

The first moment of transmissibility must be a green light for a statement to now exist. That said, do not mistake transmissibility for ontology. A statement 'exists' prior to its becoming transmissible, included in the set but not belonging as Badiou terms it. So when a statement becomes meaningful it means either you can now make that statement, or that statement will now be seen to have been made. Both are formulations of presentation becoming representation. So before a statement is determined as transmissible it must be collected within the discursive set as entirely inconsistent and meaningless. This non-transmissibility may mean no one has the courage to make the statement, but this does not mean the statement does not exist, rather it exists as void. A statement becomes transmissible when it exists, but its existence is determined by its being transmissible, or sanctioned through the axiomatic methods of a community. Yet in non-mathematical circumstances, the transmissibility of a concept is never absolutely transmissible.[37]

Full transmissibility due to mathematical axioms is part of what separates Badiou's discourse theory from his predecessors. In fact, Badiou repeatedly rejects discourse theories while stating quite freely in the introduction to *Being and Event*, that mathematics is itself a discourse (BE 8).[38] His point being that mathematics as such is discursive, but the elements it is founded on, for example the real of the void, whose truth it demonstrates, are not discursive because the conditions for mathematics are totally transmissible (recursive deduction, axiomatic reasoning). As we proceed we will replace discourse with communicability and constructability for reasons which will become apparent, most notably the term communicability allows Badiou to accept constructible worlds only if one accepts the communicability of the real of the void and the choice of the event.

As I said, communicability will come into its own in the second volume but there is a central debate in *Being and Event* between Badiou and constructivist modes of thinking of all kinds, in particular Gödel's constructivist set theory which disallows the two axioms of ZF set theory, foundation and choice, that permit Badiou to speak of a pre-discursive real element and a non-discursive new and truthful element. A second reason to raise communicability in this volume is that the event has to be decided upon due to the transmissible proofs of set theory that show you cannot rule it out but you don't have to rule it in. The totality of both volumes of the Being and Event project are, therefore, an attempt to make the event communicable and intelligible amongst you, the wider community, using modes of deduction that are transmissible to us all, such that we decide to choose the event. Being is pre-communicable, certainly, but the event is the communicability of the possibility of the non-communicable and Badiou needs our sanction for it to succeed.[39]

Theory of the subject

Badiou inaugurates *Being and Event* with a dramatic declaration of closure. An entire epoch of thought has come to an end, he tells us. The date is 1988 and naturally one might assume that this epoch designates the metaphysics of presence, deconstructed by the dominant philosophical paradigms of the day, post-existential, post-structural, phenomenological and deconstructive French thought. However, when Badiou goes on to delineate the features of global philosophy that have brought the old epoch to a halt, they are not as one might assume them to be. First he tells us that Heidegger is the last 'universally recognizable philosopher' (BE 1). Then, that the Anglo-American analytic tradition has kept alive rationality as the basis of thought. Finally, he suggests we are witnessing a post-Cartesian development of the idea of the subject. In that his contemporary philosophers of difference negate Heidegger, reject rational deductive methods and radically undermine the idea of the sovereign subject, it is clear that Badiou's new epoch is not that of deconstruction or post-structuralism. An impression further confirmed when he says that in his own work, from Heidegger he will retain the view that philosophy concerns ontology. Then, he will go along with analytic philosophy in insisting on a scientific and rational basis for thought, albeit one restricted to the mathematical revolution called set theory. Finally, he will retain the subject as the real-world praxis of the profound abstraction of ontology and mathematics. Instead of difference precedes identity, the axiom of the philosophy of difference in the ascendency around the globe as he is writing, we have ontology, mathematics and the subject. These three topics, anathema to Badiou's late-twentieth century peers, compose a neat summary of Badiou's novel philosophical direction. An epoch has ended indeed, almost before it began. In that we have already discussed how for Badiou ontology is mathematics, we are left with one leg of the tripod to teeter on, the issue of the subject, central axiom of Badiou's first truly significant work, *Theory of the Subject*, dramatically overhauled in *Being and Event* published six years later.[40]

The basis of Badiou's work is the assertion that because ontology is mathematics we can keep the subject (due to the event). To expand a little, if you can prove that ontology is mathematics then you can retain ontology as the great project of philosophy, something the twentieth century resolutely refused to allow. If you retain ontology then you can also show that, due to the nature of ontology, there can be events, moments of radical inconsistency that exceed the stability of ontology. Accept that there are events and you can have militant subjects who are committed to the implications of these events, rather than maintaining the status quo. These subjects are practical, real-world agents due to conceptual, abstract and speculative modes of thought, the kind of agents our troubled political citadel needs to admit to its citizenry, citizens which are not citizens, political subjects to a state of the entirely new.

There are profound reasons why, in the last century, both continental and the analytical schools of thought rejected ontology as the basis of thought. There also appear to be the irresolvable differences between the two strands of philosophy when it comes to the importance of rational modes of deduction in demonstrating that there are no longer truths, only meanings. And this means that until you resolve these theoretical problems, Badiou's dream of a Theory of the Subject can never be realized. The only way to unlock the potential of ontology as the basis of a new idea of the subject is to solve its logical paradoxes, but these have been in place since Parmenides and have proven to be so intransigent that, Heidegger and Russell conclude, it must be because the question itself is the problem. In other words, the game is on and Badiou is primed for the chase. He will have the subjects he desires, because of events, made plausible by a mathematical ontology, nothing less than that will satisfy him.

The quarry of the project is a truth-committed subject. This subject is possible due to a profound, nonrelational inconsistency called the event. Badiou realizes that the event is only achievable, only truly nonrelational and thus effective as a truth-maker, if the absolute consistency of being as such can be proven. Without a totally consistent theory of being, any inconsistency of events will be dubious. Badiou then schematically maps out the itinerary of the hunt for subjects in the following manner. Prove the consistency of being. Show how this consistency depends on a certain excess. Finally, tie this excess to the radical proposal of modern set theory that actual infinities exist, so that you can show that being is consistently inexistent because of the actual infinites of beings in worlds. Then, on the back of this, carve out a space for events as radical inconsistencies within totally consistent situations or worlds proven by systematic and consistent processes. Show how worlds are defined in terms of meanings while events exist because they are truths which disrupt stable meanings. Now demonstrate how events, the occurrence of inconsistent truths within consistent worlds, are only perceptible and existential in terms of their effects in worlds. These effects need to be patiently facilitated and traced. Call the non-sovereign non-conscious often non-singular not-necessarily human agents who show fidelity to events through the tracing of their truth in the situation we find ourselves in such that the consistency of that situation is undermined, subjects. Draw a bead on your prey and let loose the philosophical arrow one last time as this time the arrow will arrive, and Zeno will be confounded once and for all.

A more straightforward way of expressing this is that Badiou wants a post-Marxist and post-Lacanian, practically effective theory of the subject. He realizes that he can achieve this if he can prove that there are events. However, to prove that events occur, he must demonstrate first that beings in the world exist consistently as this is the only means by which absolutely inconsistent, because nonrelational, events can occur. Therefore, he needs a theory of being that is unquestionably consistent in its formal transmissibility. This has never been possible in the history of Western thought because all attempts to prove Being is consistent as transcendental base for multiple beings in the world have ended up showing that Being, the science of what is, cannot be, is-not, in-exists. If, Badiou reasons, he can take the impossible proposition being is-not and prove it to be rationally demonstrable as a possible and true axiom, then he can indulge his wildest dreams of what Hallward famously called 'a Subject to Truth'.[41]

The first part of *Being and Event*, then, is concerned with the basic proposition that ontology can be a consistent science if you take the result being is-not as positive by redefining the terms in play, being, the one and the many, such that being is no longer foundational of the one and the many but defined by the inter-reliance of the multiple and the void. The second part takes up this ontological consistency as a means of finally proposing a theory of radical difference that does not succumb to the simple aporia of all philosophies of difference: to define an element as different it must be relationally different from elements which are identity-consistent such that the very difference of our differentiated element is watered down, eventually to be lost. Can there be, he asks, a change that is truly different, an event, which can result in a credible post-Cartesian theory of the subject? Yes, he argues if there is a consistent theory of being due to a mathematically-founded ontology, a consistency we argue, that is only possible due to Badiou's extensive use of indifference in developing his ontology. Whereas, in the past, philosophies of difference were proposed through the negation of indifference, Badiou's theory of the event is launched on the back of possibilities for the identity-difference dialectic due to indifference, or at least that is what we will try to make communicable to you by the end of this first book.

Part One

Indifferent Being

1

Being: The One and the Multiple

Badiou launches the first meditation of *Being and Event* by posing a question that has hummed down the hallways of the Academy of Philosophy since Parmenides: How can one resolve the formal aporias or logical impossibilities of an insistence on a foundational relationship between the One and the many?[1]

Badiou contends that since the Greeks, up to the present time, including metaphysical, continental and analytical schools of thought, every philosophical system has insisted on the inter-relation of the One and the many, and based its presuppositions on this relation. In doing so every philosophical system has come across what Badiou calls the 'impasse'.[2] The impasse can be expressed by the basic formula: every foundational presupposition presents its coherence through a dependence on the assumed inter-relation of the One and the many such that it becomes incoherent. Badiou's forwardness is not due to this observation. One can find the same point in the work of Derrida, Agamben and Laruelle to name but three significant interlocutors of Badiou's philosophy. Each of these thinkers, all at odds with Badiou's ideas in some way, share his scepticism as regards the history of metaphysics' dependency on the one-many relational copula. But what Badiou refuses to bestow upon these other thinkers is his own observation that the problem lies within how we take the terms 'One' and 'many' to mean. Other thinkers tend to state that it is the relationality itself assumed to exist between One and many that is at fault, and hence work to reject this relation and move on to other, non-relational modes of reasoning. Badiou does this as well when he takes the event as nonrelational, but in terms of ontology he is unique amongst his peers in that he does not criticize relation as such or question Being as foundational. In fact, the simple formula relation + Being constitutes the large majority of all comments made across the two volumes of *Being and Event*. Instead, Badiou attempts to recompose the elements of the relation and redeem the two central components of Western thinking so far: a foundational Being due to a dialectical relation between one and many.

How to make intelligible the impossible proposition one is-not (Meditation One)

The rest of the first half of the book takes on the immense task of proving that a relational ontology of the one and the many is formally consistent. Although the road is long with many a winding turn, in a sense the whole journey is encapsulated in the first page with the central axiom of Badiou's work: 'the one *is not*' (BE 23). There is no more important phrase or axiom for Badiou than this designation, which we will write from now on as 'is-not' to differentiate this ontological observation from the more traditional uses of 'is', which are so problematic for both continental and analytical thinkers. *The one is-not* is not only the basis for the first volume of *Being and Event* but, represented as the inexistence of the whole, is also the most central statement of *Logics of Worlds*.[3] To prove that the one is-not, you first need to accept that: 'The reciprocity of the one and being is certainly the inaugural axiom of philosophy' (BE 23). If you cannot do so that is fine, but this means that in terms of the intelligibility, communicability and transmissibility of the problems of philosophy, you are refusing to participate in this world and as such you are not a philosopher. This is effectively what Laruelle decides to promote in his aggressive study *The Anti-Badiou*; refusing to play the game of Badiou, which Badiou will always win, and instead criticizing the very staging of the game.[4] I accept that this is a legitimate yet also problematic stance. In that we will concern ourselves with the communicability of Badiou's work in the second volume, for now we will accept this basic tenet citing the entire history of philosophy as our justification with almost no exceptions. In that we can cite this tradition, and this tradition composes philosophy, we can easily agree that the transmissible intelligibility of the world of objects we call philosophy is dominated by this inaugural axiom and we can proceed.

Accepting this axiom, the reciprocity of being and one as basis of all thought, then after Parmenides, right up to the modern age, you are forced into the impossible position of saying being is articulated by the differential relation of the One and the many, which becomes exceptionally problematic if you designate that either the One or the many 'is'. It is very likely, given the way the problem is posed, that you will want to say that the One or the many 'is', after all ontology is the science of what is, and if you do that it is inevitable that you will be forced to deduce that if one of the elements in this reciprocal relation 'is', then the other, which is different, is forced into the position of being the 'is-not'. This is indeed what happens time and again across the history of Western thought until the beginning of last century.

For two and a half millennia, the greatest minds of our worlds have accepted the axiom of the reciprocity of Being and One, and yet failed to solve the logical problems this has resulted in: if the one 'is', then the many is-not, but if the many 'is', then the one is-not one. Badiou, however, is nothing if not an optimist. Inspired and supported by the formal axioms of set theoretical mathematics, he presents a way in which ontology and relation can be retained and the intellectual project of the West redeemed. If you allow, he says, that ontology is based on the axiom that the one is-not, that its existential presence is that of a void, and you accept that instead of the One you are speaking of something that counts-as-one but is not One (not a transcendental whole) then you enter a clearing in the thicket that has so hindered your pursuit. Enter that clearing, he implores you, and that problematically inconsistent many can be conceived as what set theory takes as a pure multiple and all the logical impossibilities of the system will be removed. In other words, if you combine mathematical axiomatic reasoning with a functional rather than referential or expressive or conception of terms like 'being', 'one', 'many' and 'is',[5] then you can retain the one + many relation as the basis for your definition of being. If you do this, the idea of philosophy, the reciprocity of being and the one, remains communicable at the precise point in its history where this seemed no longer possible, the schism between the philosophers of difference and those of formal constructability. It is a great story, the greatest ever told now that God is dead and wizards have faded into decrepitude, but for it to be more than just a yarn we tell in the woods at our epochal twilight, Badiou is compelled to be able to demonstrate it in a manner that is communicably transmissible amongst us, all of us. This is what the rest of the book aims to achieve by combining ontology, the continental tradition, with logic, the analytic tradition, through a philosophical reading of the mathematics of set theory. Put together the methods of ontology, formal constructability and mathematics, he argues, and we are back on track, ontologically speaking.

The one as operational count-as-one

The first meditation is concerned with the one and the multiple as *a priori* conditions for ontology. More widely it delineates the various basic concepts that Badiou will require the reader to accept to best understand how the one and the multiple compose the field of a renovated and effective ontology. Having presented the apparent paradox that *being is-not,* Badiou explains how this is possible if we take the one to be simply an operational count-as-one, not a unified transcendental totality of all predicates included in an object or all objects contained beneath a single defining heading. The operational one is a method that he takes from the mathematics of set theory. The particular power of set theory is that it allows Badiou to state that the one is-not in such a way as not only is this not a problem, it is in fact a constructive solution. In set theory, a central axiom is that in any presented situation the one of that situation, what allows us to say it is a single consistent situation, counts-as-one but does not compose *the* One. Due to this, the many in any situation is in fact the abstract presentation of any multiple whatsoever, not a specific multiple possessive of certain qualities. We will consider the axioms of set theory soon enough but I strongly believe that in Badiou the mathematics of set theory is simply a more formal way of presenting ideas that already exist in philosophy or which can be easily expressed therein. So, leaving the set theory specifics to the side for the moment, let us concentrate on the basic idea of the one as an operation not a thing, and then the many as quality neutral or what we will call indifferent, as these are the two most important philosophical ideas Badiou presents though his analysis of sets.

To recap, when you speak of the one of many things, you are speaking of the operational counts-as-one, not a transcendental essence that all these things share in common such that we can say they are all particular examples of chairs, cats, unicorns and so on. In addition, when you consider the nature of each of these counted ones, they are merely indifferent multiples definable only in that they have been counted. They are not actual objects with definable qualities or predicates. If your set is called cats, in set theory, the cats in that set are 'cats' only in as much as you have counted them to be cats. If you are speaking of the one as a collective one, you are merely speaking the operational consistency of elements as countable as ones. If you are speaking of the one as a single one, you are actually referring to a countable one as defined simply by the fact that it can be and has been in this instance counted as one along with other ones.

We will come back to the indifferent nature of single ones later. For now, we will satisfy ourselves with analysing the operational nature of the one, rather than its thingly status which is what is usually emphasized. Of this operational nature of every one, Badiou says:

> What has to be declared is that the one, which is not, solely exists as *operation*. In other words: there is no one, only the count-as-one. The one, being an operation,

> is never a presentation. It should be taken quite seriously that the 'one' is a number . . . In sum: the multiple is the regime of presentation; the one, in respect to presentation, is an operational result; being is what presents (itself). On this basis, being is neither one (because only presentation itself is pertinent to the count-as-one), nor multiple (because the multiple is *solely* the regime of presentation) (BE 24).

What does it mean here first that the one is an operation, and how does this show that the one is-not? The operational one is what is produced from the need to gather together the elements that are present in a situation so that you can say these are the elements present in this situation. The elements are individual units, you can count each as one, and if you can do so then you can also count all the units present as an overall one as well. For any situation all you need to be able to say is that this is a situation, it is so because these are the elements that compose it, and finally these elements that compose it are the presented elements of this situation.

Although set theory remains intimidating to many, in fact its basic assumption, which we have just described, is brilliantly simple and profoundly philosophical. A set is an aggregation of elements. Or if you prefer: A set is a collection of thing*s*. In effect, what you are saying is that: Every one is an indifferent operation. All a one does is gather together other ones, this is the operation, such that all you can say is that this is the count-as-one of all these ones, this is the quality neutral or indifferent nature of the operation. What set theory allows you to do is gather together any number of neutral 'ones' into something you can count-as-one. This means in practice that in set theory you always have two counts. You count all the elements that are present as 'ones', namely the things you are dealing with. This is called belonging or presentation. Then you count them again as a collection of these ones. This is termed inclusion or representation. Set theory is encapsulated, philosophically at least, in the nature of these two counts, presentation and representation, and their inter-relation.

The ancient problem of classes

It is important in set theory to understand the particular relation that the second set, the count-as-one, has to the ancient problem of classes, as here the two modes of reasoning, mathematical and metaphysical, come together. Many thinkers have insisted that class precedes membership. In this way of thinking, the whole precedes the parts and allows you to then say, when a part is presented, this is a part of this whole or this is not a part of this whole. As Mary Tiles shows in *The Philosophy of Set Theory*, the issue of classes has tormented philosophy since its inception with the work of the Greeks, in particular Aristotle.[6] That's because it relies on the ancient aporia of a foundation dependant on a relation. While it seems intuitive to think of the essence of something, its class, as existing in advance of counting that thing – how can I count this as a chair if I don't first have the concept chair – for set theory the counter-intuitive position is central. In set theory, the statement 'this is a chair' is possible only because it is designated in the counts-as-a-chair set. Yet if the second, classificatory count, always comes 'after' the first presentative count, I can only class this a chair because it is represented in the set Things which we Can Count as Chairs, then the elements that are presented as ones to be counted are only presented as such due to the second count. In this way, in some sense, the first count, which is always first, comes after the second count, which however is always second. If, after Badiou, we call the first count presentation and the second representation, we can propose the following operational formula. In any count, representation occurs before presentation. This contradictory stance is only possible if we accept that the one is an indifferent operation, not a differentiated thing, something set theory is entirely at ease with.

If we think of the one as an indifferent operation rather than a thing as such or an essence of a thing as such, the two traditional ways of thinking of a one, we can say the following. Every object presented is only presented when it is presented in or to a situation. Even if no situation can occur until some elements are presented. There are no objects away from a situation, and there is no situation devoid of objects. For reasons we will go into later, even a situation of no objects contains at least one element, the void set. The one then 'exists' solely as an operation that allows us to count, both to count objects or elements as elements, and to count a collection or aggregate of elements as compound elements. One can also see here that said aggregate itself could then be an element of a larger count, meaning that every aggregation, up to a point, can be counted either as one element, or as an element of a set. Every aggregation, up to a point, is both a set and a multiple of another set. All that exists are sets of sets or as Badiou prefers multiples of multiples.

Coming back to the history of philosophy, I hope the reader can see that this basic tenet of set theory is, as Tiles explains, merely restating the age-old problem of classes. What matters for us, and I am taking myself to be resolutely a philosopher and not a mathematician on every page of this book, is not so much the local, situational mode of counting, this is relatively easy to present. Rather, the issue that worries us is to be found precisely in that 'up to a point' that I mentioned repeatedly. Perhaps it would be better to say 'down to and up to a point'. The problem of aggregation is, after all, nothing but *the* problem of philosophical thought. If every element can be simultaneously a one and a one-of, then the problems of infinite regress and bad infinity rapidly arise. Badiou must resolve both these problems to allow his contention to remain.

Situations and structures

Badiou decides early on to fix the terminology of his ontology. A situation, for example, is 'any presented multiplicity' (BE 24). In *Logics of Worlds*, the term 'situation' will permanently be replaced by that of 'world', or any presented set of existing relations in terms of their appearing in a world. We will keep situation on the whole in this first volume to make things clearer for the reader. The word 'any' is important here. Not only is a multiplicity indifferent in terms of its being quality neutral, but its presentation as far as ontology is concerned is also indifferent, this is what we take 'any' to signify here. Badiou confirms this by saying a situation is the place of a taking-place, 'whatever the terms of the multiplicity in question' (BE 24). The terms, the qualities of a multiplicity, are entirely irrelevant, so a situation is fundamentally without a specific term and in this sense it is class, term, quality and predicate indifferent or neutral. This is perhaps one of the most radical and initially confusing elements of Badiou's immanence. Every presented one is presented in a situation and so is immanent. It is a being for a situation. Yet this immanence is absolutely abstract, contravening the usual motivation for immanent thought which is localized specificity. A one is in the world yet it is entirely neutral in terms of its qualities at least when it comes to its being.

Even more difficult in this regard is his insistence that 'Every situation admits its own particular operator of the count-as-one', or what he calls structure (BE 24). Confusingly, the operational nature of the one, which appears neutral, is described here as being particular to each situation, which seems contradictory. This is perhaps best explained as regards the idea of a cardinal number in set theory. A cardinal number determines the specific size of a set. Each set has only one cardinal number which coincides with how many elements there are in that set: 1, 2, 3, 5, infinity . . . Yet if one were to find two sets containing different elements but of the same size then their cardinalities would be the same and so they would be the same sets. Cardinality requires the reader to rethink what they thought they knew about sets perhaps gleaned from drowsy afternoons at school. Instead of thinking of a set as a set of X, say a set of cats or avant-garde poetic techniques, you need to think of sets as a set of X elements or of X 'size'. The result of this is that the count-as-one is an indifferent operation, but it is not abstract because the operation only exists for a situation, you only count if there is something to count for, and because each situation has its own count or structure: 'the regime of its count-as-one' (BE 24). If a set is called 5, then it is particular. It is not 4 and it is not 6. However, this set is still indifferent in that you call it 5 because it contains 5 elements, but you have no idea what the nature of those 5 elements actually is or rather you don't need to know anything about their nature to define them as belonging to the set 5.

Badiou now enters into the crux of the matter as regards the axiomatic method typical of set theory. In terms of being-counted he notes: 'When anything is counted as one in a situation, all this means is that it belongs to the situation in the mode particular to the effects of the situation's structure. A structure allows number to occur within the presented multiple' (BE 24). To count something in a situation only means it can be counted as one due to the structure of the situation. Yet this structure only exists as an operational means of allowing something to be counted. To be clear, we have to accept the circularity of there being no count without a multiple situation, and no multiple situation without a count as this is nothing other than restating the law of the set in set theory. If, as Badiou continues, 'a structure allows number to occur within the presented multiple' (BE 24), then, he asks, does this infer that the multiple is not yet a number? If this were to be the case the house of cards would collapse for Badiou as spectacularly as it has for other thinkers tasked with this causal-successive paradox in that the foundation, number, would proceed from the particular example of that foundation, the multiple. Here then is Badiou's answer upon which, I believe, much else depends as regards axiomatic retrospective reasoning:

> One must not forget that every situation is structured. The multiple is retroactively legible therein as *anterior* to the one, insofar as the count-as-one is always a *result*. The fact that the one is an operation allows us to say that the domain of the operation is not one (for the one *is not*), and that therefore this domain is multiple; since, *within presentation*, what is not one is necessarily multiple. In other words, the count-as-one (the structure) installs the universal pertinence of the one/multiple couple for any situation (BE24).

There is a dark brilliance here. Badiou takes the precise aporia that has been the impasse of all thought, namely the problem of the One and the many, and in uncoupling being from the One he is able to use the paradox of the One and the many as constructive basis for this proof.

This is typical of axiomatic thinking. Take being. Instead of thinking that being is One, presented through the multiple, assume that being is-not or that its consistency is-not. This is unthinkable because ontology is the science of what is. If you say that being is all that is-not you say something impossible. Yet, if you are willing to say this, then the one and the multiple cannot merely live in harmony, but more than this they co-found each other and in doing so prove your original hypothesis, namely that being, uncoupled from the One, is-not. This astonishing form of axiomatic retrospective proof is so workaday for mathematicians that it almost goes without comment.[7] It is also, of course, the basis of twentieth century phenomenology in the hermeneutic circle. Finally, it is another way of stating the basic formal law of constructability. It is time, in other words, that retroactive axiomatic reasoning become fully communicable amongst us in all the work we, as a community, go on to produce.

The multiple

Every one, as we saw, is an operational or count-as-one, not an actual, unified, predicate-possessive thing. In addition, we can only know of the being of a one when it is presented to a situation. A situation, by definition, is structured and structure is an order relation between at least two elements. Therefore, it is reasonable to say every count-as-one counts as one in a structured situation composed of at least two ones or every count-as-one is well ordered. Thus every one is also at the same time a multiple. As Badiou says, although the multiple is retroactively legible and so naturally comes after the one – how can there be a set if there is not at least a one – in that every one is a count-as-one, not an actual, existing one object (it can't be as the one as such is-not) in fact it is to be read as coming first. The count-as-one is always a result, naturally, and so something must come before to facilitate the result. That something is the multiple of the situation or world. More than this, if the one is an operation, then the 'domain' of the operation, what allows you to count-as-one a one, cannot itself be a one, not only because the one is-not but because the one can only be counted as a result due to a situation. This proves that the domain of the count, which exists because it is presentation, in that it cannot be one, can only be multiple. As he says, in any presentation, whatever is not one is a multiple because within any structure all you have are two existential possibilities: that something is a one and that this one can be counted. To allow for the apparent paradox that something is a one because it can be counted and something can be counted because it is a one, Badiou proffers the formula that being as such is-not.

At the end of this section in a rather compressed sentence Badiou says: 'What will have been counted as one, on the basis of not having been one, turns out to be multiple' (BE 24). We have three factors in play here: the not-one, the one, and the counted-as-one; or if you prefer being, count-as-one, and multiple. If you uncouple being from the one, or the many, and place the terms in a triangular inter-relationship where each element depends on its other two then that, in a nutshell, is the basis of Badiou's first conception of the ontology of the one. This model benefits from clarity but it lacks something in terms of accuracy as Badiou immediately explains when he presents an, essentially, bifurcated version of the multiple. Central to the retrospective axiomatic method he is proposing here, is the fact that the multiple is a part of two elements of our triangle: '"multiple" is indeed said of presentation, in that it is retrospectively apprehended as non-one as soon as being-one is a result. Yet "multiple" is also said of the composition of the count, that is, the multiple as 'several ones' counted by the action of the structure' (BE 25).[8] These two multiples, of inertia and of composition, Badiou decides to call inconsistent and consistent multiplicity. From this he is then able to say that a situation (structured presentation or world) is defined by double multiplicity. This double multiplicity is also placed in a clear order of succession, inconsistent multiple first, consistent after. Yet, at the same time 'Structure is both what obliges us to consider, via retroaction, that presentation is a multiple (inconsistent) and what authorizes us, via anticipation, to compose the terms of the presentation as units of a multiple (consistent)' (BE 25). All of this seems the result of two ways of looking at or cutting the same thing. So 'it is the same thing to say of the one that it is not, and to say the one is the law of the multiple, in the double sense of being what constrains the multiple to manifest itself as such, and what rules its structured composition' (BE 25). We will hold off for now defining what the multiple actually is, as Badiou is careful not to overuse set theory terms until the philosophical development of his ideas is firmly in place.

Presentation of presentation

Badiou's radical immanence is captured in the axiom 'There is nothing apart from situations. Ontology, if it exists, is *a* situation' (BE 25). What this demands of us is rather than think of how being makes operative a situation, we have to work out retroactively how a situation makes 'possible' ontology at all. For, as Badiou says, if a situation is presentation, how can being be present *qua* being if, by definition, it is-not and so cannot be presented in a situation? Remember, to be present in a situation you need to be available for the count and being as a whole, as such or as a thing, is never available for the count. Further, if being is a situation, then it must also submit to structure through collective counting. Yet being is subtracted from every count as that which allows for the count but which is not counted. By definition, Badiou concludes from this, 'there is no structure of being' (BE 26).

There are, Badiou notes, various traditions that share similar designations for a being that 'cannot be signified within a structured multiple', and that 'only an experience beyond all structure will afford us an access' (BE 26). The first of these is Platonism, keep in mind that Badiou is an unashamed Platonist in terms of the realism of an object beyond mathematical constructions, resulting in a negative ontology of being. These have led to the designation of being as the absolute Other, being as in a state of withdrawal, and being as poiesis (because only poetic language is capable of presenting exceptions to situations).[9] He calls these ontologies of presence and, along with Derrida, implicates Heidegger especially in this regard before rejecting them outright.

If being cannot be structured and yet is not captured by ontologies of presence then how does being present itself such that we can say it exists? Badiou says, brilliantly, using the entire problematics of the deconstruction of the philosophy of presence which typifies the tradition of the philosophies of difference: 'If there cannot be *a* presentation *of* being because being occurs in every presentation – and this is why it does not present *itself* – then there is one solution left for us: that the ontological situation be *the presentation of presentation*'' (BE 27). As this is a complex formulation and one with wide-reaching implications we should commence with a simple description. Let us say that being is the presentation of presentation if we say that every time there is a presentation, being is not presented but is 'present' as that which must be subtracted from every presentation for presentation as such to occur. So, every time you note a presentation you are also noting being by just remarking that there is a presentation, because being is that which cannot be and so is not presented in every presentation. There is quite a bit to do to prove this, but to accept it all you need to concede is that there is some thing rather than nothing, better expressed as there is some thing due to no thing. If you can accept that there is some thing, simply some thing presented with no other value judgement here except to say that it is presented, then you can admit to Badiou's ontology of the multiple that depends on the subtraction of being.

We now have to burrow down into the multiple as ontological presentation of presentation. Assuming we accept that some thing can be presented, and that this does not disallow being, which is not presented, and that being is-not allows for the presentation, which we facilitated by uncoupling being from the one, we still have a lot to contend with. Badiou starts by explaining that the only predicate we have applied to a situation is the multiple. If the one is not 'reciprocal with being' i.e. the one is totally nonrelational as regards being, we can say that the multiple is reciprocal with presentation if we accept that multiplicity is composed of inconsistent and consistent modes. Consistent multiplicity is not the problem, specific presentation is fairly straightforward, but in terms of the presentation of presentation we are speaking in general terms about presentation as such. This is essential, ontologically, but also problematic. Badiou explains as follows:

> Presentation 'in general' is more latent on the side of inconsistent multiplicity. The latter allows, within the retroaction of the count, a kind of inert irreducibility of the presented-multiple to appear . . . if an ontology is possible, that is, a presentation of presentation, then it is the situation of the pure multiple, of the multiple 'in-itself' . . . ontology can be solely *the theory of inconsistent multiplicities as such*. 'As such' means . . . without any other predicate than its multiplicity (BE 28).

In place of presentation in general, we will speak of indifferent presentation as regards the presentation of presentation. This makes sense in relation to Badiou's comments. Presentation means can be counted. As we saw, ones can be counted only if they are indifferent, or, as he says, if their only predicate is that they are multiple. This is not really a traditional predicate of course; it is an operational feature which is applied retroactively. Being a multiple as such is a predicate that occurs before the object, and it is a predicate that is not a quality of the object because all it says is there is an ob-ject. When Badiou speaks of ontology as the theory of inconsistent multiplicities as such, what he actually means is that ontology is the indifference of pure presentation. Something can be present, and that is all you can or need to say about it. Yet we ought to be clear that this something that is present is determined as present simply because it can be counted as a potential. That said, it has not yet *been* counted. It is in the latent, inert state, but something about it means it must be able to be presented. Again this takes us back the importance of operativity over quality when it comes to being qua being. Being is the operational basis for presentation. For some thing to be present, being must be present as absent.

Badiou now makes a crucial distinction when he asks what the presentation of presentation could be composed of in that it cannot be made up of ones as the one is a result of presentation not an element of presentation. Ones occur because there is a presentation of presentation, so if presentation is not composed of ones, then it makes sense to say that it must partake of the same quality as being. If being is-not, he argues, then presentation is to be defined as the 'without-one' (BE 28). Yet if this is the case, as presentation must come to presentation, whereas being does not have to (it is a defining feature of presentation that ones will be presented) what can one define as the precise nature of the composition of presentation, determined as it is as being without-one?

First, the multiple from which a situation is produced is itself made up entirely of multiplicities. 'There is no one. In other words, every multiple is a multiple of multiples' (BE 29). It is important here to realize how distinct a multiple is from a traditional sense of multiple as the many. This is an entirely indifferent multiplicity of which all one can say is that it is multiple and therefore is-not one. And of course we are all sensitive to the infinite proliferation implied in the phrase 'multiple of multiples'. For this to work in any sense as a formulation we need strong measures to deal with the infinity implied here and the infinite regress at its base as well. The second point is that the count-as-one that determines the first presentation via the first count is not facilitated through the pure multiple as such, but rather the other way around: 'The count-as-one is no more than a system of conditions through which the multiple can be recognized as multiple' (BE 29). The whole point of the first count is to present the ontological system of inconsistent multiples. It is because multiples can be counted and thus structured into a consistency, that we can prove that there are such things as inconsistent multiplicities or multiples of multiples or, for our own system, presentations of presentation.

The concept of inconsistent multiplicity is the key because it proves two central features of Badiou's ontology and his valorization of the disruption of ontology due to events. The first is that consistency is just a means of proving inconsistency and the in-existence of being. This is the prize. He will retain the right to speak of structure in another fashion for the second volume of the work. The second is that the means by which the count presents presentation as pure multiplicity, shows that the event is either potential in every situation or, if that is too relational, every situation contains at the very least an example of nonrelationality intrinsic to a proper understanding of the event if and when it occurs. However, the second point strikes Badiou as too extreme because it leads inevitably to the sense in which ontology can only conceive of multiplicity in terms of a multiple, which I hope is not what we have said but is certainly there as a threat. Naturally, if presentation is presented only through the count retroactively, and the count is nothing other than counting a multiple as one, pure presentation runs the risk of being counted in some way. Instead of this, he argues, 'What is required is that the operational structure of ontology discern the multiple without having to make a one out of it, and therefore without possessing a definition of the multiple ...' (BE 29). To achieve this, something about the count-as-one must accept that it can only count pure multiples. Any other options 'whether it be the multiple of this or that, or the multiple of ones, or the form of the one itself" (BE 29), are always already in the process of making multiplicity *a* multiplicity of some sort. This immediately adds definition to the multiple but the multiple, which is indifferent, is therefore also in-definite. Badiou finally explains that if presentation cannot say 'I accept only pure multiplicity', for this would add definition to pure multiplicity (even our own stipulation is excluded here in that indifferent multiplicity is still a *kind* of multiplicity) then this prescription, as he calls, it must be implicit. It is not a stated intention but a tangible result of the count. The final piece of the puzzle falls into place at this juncture: the presentation of presentation is nothing other than the axiom.[10]

Reasoning on being by means of axioms

The presentation of presentation can only ever be axiomatic. Although we might speak of pure presentation, presentation *qua* presentation, indifferent presentation, multiples of multiples and so on, these descriptive phrases are fundamentally insufficient. All of them participate in a metaphysics of predication which is logical. And ontology is not logical. In as much, as Badiou says, that logic is a system of relations, there can be no logic of being as being is by definition indifferently nonrelational. If this is true, then all of the above constructions are retroactive presentations of presentation as a thing. This is not what is meant by the presentation of presentation as Badiou makes explicit using axiomatic mathematic reasoning:

> What is a law whose objects are implicit? A prescription which does not name – in its very operation – that alone to which it tolerates application? It is evidently a system of axioms. An axiomatic presentation consists, on the basis of non-defined terms, in prescribing the rule for their manipulation ... It is clear that only an axiom system can structure a situation in which what is presented is presentation. It alone avoids having to make a one out of the multiple, leaving the latter as what is implicit in the regulated consequences through which it manifests itself as multiple (BE 29–30).

In an axiomatic system, if it works then it is 'there' without it having to be counted because it never has to be explicitly stated. The same is true of language. Language, as Heidegger famously says, never has the floor. We can say it exists only due to its effects. This is why the old adage that you can only ever say what language is through language is effectively a false version of Russell's paradox. Language is not a thing but an operative axiom and in as much as mathematics is a non-referential language it simply makes this easier to see because the false quality of reference, key to the foundational sense-reference couple of logicism, is obviated. Mathematic symbols do not refer, they facilitate an operation, meaning you can count as a process without counting actual things. This is why our contention that Badiou's system is unthinkable without indifference is so strong. If ontology is mathematics then, by definition, ontology is indifferent because mathematic processes, axioms, are indifferent by which read purely abstract in being non-referential and without sense.

Badiou's concluding comments on the first meditation reveal the centrality of the axiomatic system to the presentation of presentation upon which the ontology of being depends. He begins by reiterating that the axiom of being is that it is an inconsistent multiplicity. By this we mean that being is a multiplicity that is not yet counted, for counting is what produces consistency through the presentation of being within a structure. That said, the effect of the axiomatic system is to make inconsistent multiplicity consistent through deployment, specifically this is the effect of the presentation of presentation. In so doing, inconsistent multiplicity is presented without it becoming a one, a unified whole. Badiou says here of the presented one that 'It is therefore absolutely specific' (BE 30). Inasmuch as counting-as-one is indifferent, the one is merely present to be counted but has no defining qualities as such, the presentation of presentation is, perhaps confusingly, resolutely particular.

We now enter into a complicated consideration of particularity and indifference. When pure multiplicity is used operationally it is entirely devoid of qualities. In this way it is fundamentally indifferent. Sets and their components are abstract; they are not classes of things but operational aggregates of elements. Before the axiomatic operation typical of ontology, the elements in play in-consist due to the fact that they are ones and particulars. Badiou calls this the impure multiplicity of a class of named objects. Since the Greeks we have known that classes of actual things present an unsolvable aporia, which is that, in this sense, they in-consist. Axiomatic theory however can treat objects that are inconsistent in a consistent manner. While normal class-based ideas of sets treat objects as consistencies but, in doing so, find a fundamental, operational inconsistency. Or as Badiou puts it: 'To accede axiomatically to the presentation of their presentation, these consistent multiples of particular presentations, once purified of all particularity – thus seized before the count-as-one of the situation in which they are presented – must no longer possess any other consistency than that of their pure multiplicity, that is, their mode of inconsistency within situations' (BE 30). This is a detailed way of saying that for objects to be counted they must cease to be 'objects to be counted' and become, instead, abstractions within an operation. The only way to make particularity consistent is to treat it as a real particularity by which we mean a very specific result of an axiomatically sound set of operations. Particularity here then is not an object with its predicates but a multiplicity within an operation.

The result of taking the one as a multiplicity within an operation not an object with qualities is that in the axiomatic system, primitive consistent particularity, this is chair it has four legs, is inconsistent. Also in philosophy since the Greeks. Yet inconsistency taken as '(. . . pure presentative multiplicity) is *authorized* as ontologically consistent' (BE 30). Authorized through consistent use within the axioms of reasoning. The choice is between consistent particularity that is ontologically inconsistent, and inconsistent particularity which is ontologically consistent. Badiou chooses the latter with the idea that in terms of particularity, indifferent particularity, the details can be added in later.

He concludes on this rather astonishing set of propositions with a rare, positive nod to Derrida when he says:

> Ontology, axiom system of the particular inconsistency of multiplicities, seizes the in-itself of the multiple by forming into consistency all inconsistency and forming into inconsistency all consistency. It thereby deconstructs any one-effect; it is faithful to the non-being of the one, so as to unfold, without explicit nomination, the regulated game of the multiple such that is it none other than the absolute form of presentation, thus the mode in which being proposes itself to any access (BE 30).

In a masterpiece of brevity, the totality of Badiou's theory of being is contained in this conclusion. Ontology works through axiomatic operations. These operations can be accepted by a community as absolutely transmissible due to their operative efficacy without ever having to answer ontological questions such as Do Sets Exist? What is an Aggregate? and so on. This is how mathematicians operate. A second feature here is that not only do we not need to prove the ontology of the axioms, we do not need to prove their consistency outside of use, and we do not need primitive consistencies, actual objects collected by concepts, to use them. Axioms are non-referential, require no meta-mathematical proof beyond their efficacy, and work entirely with inconsistent multiplicities: *x*, *y*, *a*, *b* etc. This, Badiou argues, is the true way to deconstruct the one-effect which is best defined axiomatically in negation, as befits Badiou's subtractive ontology.

Axiomatic reasoning of this order means if you take the one as non-being, you can count multiples as entirely indifferent in terms of quality. If you do that you make them completely particular in terms of context of usage. This allows one to present presentation without counting it in any way as some thing. Such a form of entirely indifferent, abstract presentation allows one to have access to the one as-not which was our original axiomatic proposition. Put simply: if you assume the one is-not, you can operationally prove the one is-not through the presentation of presentation in which the one can be present and yet not 'be'. This, in miniature, is Badiou's ontology, his methodology (axiom systems) and his epistemology (how impossible propositions become consistent truths). We now have our basic axiomatic proposition: being is-not. Now we need to retroactively prove that this absolutely is the case.

2

Being: Separation, Void, Mark

Meditation Two

The second meditation, 'Plato', as Norris points out, is primarily interested in taking Parmenides' dictum, 'If the one is not, nothing is' (BE 31),[1] and changing the emphasis to a productive sense that being is-not. If read in this way, as we saw, the logical impossibilities of our tradition are no longer an impasse but an opportunity.[2] Badiou reads Plato, amongst Parmenides and Socrates, as demonstrably right on four points of ontology: the one-being, the there-is of the one, the pure multiple and the structured multiple (BE 36). And lamentably wrong in one key regard, 'because the gap between the supposition of the one's being and the operation of its "there is" remains unthought' (BE 36).[3] All that is left to say is that Badiou is Platonic, up to a point. At the time of publication this was an incendiary position to take up, but with the sober reflection of three decades and a wider understanding that Platonism of one form and another is rife throughout the philosophy of mathematics, I think we can pass on from it with little more comment as regards our own study.[4] From this point on we can outline two clear directives in relation to our methodology: to understand Badiou's ontology in detail and formalize its relation to indifference. This gives us permission to only pause on those sections of the study that add something to our understanding of this ontology or reveal something more about indifference. We have Badiou's blessing in this regard where in the preface he says the book is composed as a three-in-one structure: conceptual, textual and meta-ontological (mathematical). As a rule of thumb, our study will tarry as long as is necessary on the conceptual, using the meta-ontological to form the mathematical origins of what is said there, and turn to the textual only if absolutely necessary.[5] I think we are safe in this procedure due to the excellent work others have done on Badiou's relation to the history of philosophy. And so it is that we rifle through the second meditation only so that we can buy the luxury of lingering in the pages of the third which, as it is concerned with defining the pure multiple, is an essential moment in the history of thought.

Set theory and aggregation as collection (Meditation Three)

In the third meditation we are very much in the realm of the mathematically meta-ontological and so we ought to further define set theory, specifically in terms of its role in Badiou's ontology. In fact, we have already explained a general use of set theory: it is axiomatically indifferent. The axiomatic nature of set theory means that the elements counted as collected are always without qualities other than those abstract procedural axioms necessary for set theory to operate. We can now say more, but not much more. As Potter says 'sets . . . are a sort of aggregate'.[6] As to what an aggregate is, it is perhaps comforting in a fashion to discover that this is as highly contested as say what being or truth is. Ordinary language treats a set as 'a single entity which is in some manner composed of, or formed from, some other entities' (Potter 21). The repetition of some here is significant for Norris' admission that 'sets are defined as products of the count-as-one, that is the classificatory procedure that consists in grouping together a certain range of such entities and treating them as co-members of a single assemblage whatever their otherwise diverse natures or properties . . .' Which allows the set theorist 'to ignore any merely contingent or localized differences between such entities and accord them strictly equal status as regards their membership of any given set' (BBE 50–1).

In terms of both definitions, and more generally *all* definitions of sets, the specific qualities of the entities collected are negated and replaced by more straightforward, operative qualities. This is where set theory is beautifully simple in that these qualities are reducible to belonging, inclusion and exclusion; or set, subset, and not in the set. Every set is indifferent in terms of it being the collection it is, its elements are indifferent, they either belong or are included, and indifferent in terms of everything else, which is simply not included. Even if you extend the designation of the set to include how sets interact, in terms of union, intersection and complement or making a set of two sets, making a subset of elements of two sets, and making a set of everything in the world not included in either set, indifference is still the dominant force.[7] In set theory it does not matter *what* you are but *that* you are due to *where* you are within the count combining our law of indifference and our maxim of communicability.

One last thing needs to be said about sets. There is a distinct difference in most versions of set theory, and in particular Zermelo-Fraenkel, over what it means to be an aggregate. Philosophers speak of two ways of aggregation relevant to set theory, fusion and collection. Potter clarifies: 'a fusion is no more than the sum of its parts, whereas a collection is something more . . . A collection . . . does not merely lump several objects together into one; it keeps the things distinct and is a further entity over and above them' (Potter 21–2). Why is this relevant? Well, for those readers familiar with sets and unfamiliar with set theory, one might mistake a collection of things as simply all the things there are under the class of those objects. Potter gives the example of a pack of cards. The collection of the pack of cards is precisely this pack, which is composed of fifty-two members. The fusion of the cards is rather: we took these various cards together as a set, which could be all the cards, all the suits, all bent cards, or a random selection of cards. Perhaps it is best to think of the collection of cards as the 'Pack' of cards which designates a specific number of members. Then of fusion, all the ways the cards can be cut. Membership, in other words, requires an additional order of designation, a kind of inverted commas although usually in set theory it would be represented as {pack} of cards.

The difference between set as fusion, the common presupposition, and set as collection, the actual conception of set in mathematics, will turn out to be a crucial distinction, so let us give two more examples from Potter. The first concerns sets with only one member. A collection with only one 'member' has, in fact, two elements. These two elements compose the set as such plus its member, often called a singleton. So, the collection of Mallarmé is {Mallarmé} a set to which one thing belongs. The cardinal number of this set is 2: two Mallarmé's, the {Mallarmé} collection and the single Mallarmé member of the collection. Whereas the fusion of Mallarmé is just Mallarmé, and has no members at all. It just is or rather he just is. The other example is the set with no members. Potter says a container with nothing in it is still a container, so that an empty collection is a collection with no members. The cardinal number of this set would be 1, not 0 as you might imagine. The collection is one element even if it is empty; jugs are still jugs even if they contain nothing, to use a Heideggerian-inflected example. In contrast, a fusion with no members is simply impossible. Potter says you cannot fuse when there is nothing to fuse, but perhaps it is better for our purposes to say that nothing can be included in a set to which nothing belongs or again, there can be no subset of a set with no members. To clarify, there is a subset of a set with no members because membership, the set of one object, and inclusion, the one object in that set, are operationally different.

On the back of these considerations set theory can be defined as the process of collecting based on one simple, non-logical primitive: the clear division between there being such-and-such a set due to a determinate number of members, and what can be collected within that collection. The key things to remember are this sense of collection rather than fusion, or a set is a self-conscious counting of determinate elements, that these elements are quality neutral, they are nothing more than the elements collected, that due to this there can be a set with only one member, meaningless if you are talking about an object per se, and there can be a set with no members, impossible if you are speaking of actual objects in the world. Finally, due to the idea of sets as collections in terms of belonging, the possibility of sub-sets is presented, an essential ontological stipulation for Badiou as we shall see.

All that set theory actually does is make abstract or indifferent the items collected, and separate out two levels to belonging. If you accept a set as indifferent and bifurcated, Badiou argues, you solve the impasse of ontology: the idea of being as a foundation due to the relation between a one and a many. So, whenever we consider set theory from now on, it is to allow this idea of ontology to remain communicable amongst us. That said, a second use is also important to Badiou, specifically how set theory facilitates a totally nonrelational theory of the event.

Axiom of separation

Although Meditation Three is ostensibly about the theory of pure multiple and its paradoxes, it is more accurate to say that it is primarily concerned with separation. In particular, the axiom of separation which forms a central part of ZF. The axiom of separation is a formal proof of the basic nature of the multiple which is that it is both inconsistent and consistent. When a multiple is said to belong to a situation, it is inconsistent. By that we mean we can count it as a one in a situation, but we do not yet know anything about the structural presentation of that one relative to the overall situation. When a multiple is said to be included in a situation, we are able to say something about the multiple relative to other multiples that belong to the same situation. For example, we can say that all the ones are now counted-as-ones as parts of the overall set. Or we can say these three ones are counted as a subset of ones which is itself a one of part of the overall set. In other words, if all there is are multiples of multiples due to the fact that the one in-exists in every multiple, then each multiple has a double aspect. A multiple is both a one as such, belonging to any situation, and a count-as-one, included in this situation. Every multiple is both a single unit, and a part of at least one other unit. So each multiple belongs and is included at the same time. A one can either be an indifferent unit, a single one, or an operational count, count-as-one, depending on how you look at it. You could also say this is the difference between multiples and sets.

Returning to set theory, the axiom of separation formalizes this basic quality of every multiple as both inconsistent, belonging, and consistent, included, or as indifferent one and operational count.[8] As this is one of the two foundation stones of Badiou's entire ontology, the other is that being is-not, it is important that we accept the formal proof of separation if we are to stay with Badiou in all that is to follow. What separation says specifically is 'given a multiple ... there exists a sub-multiple of terms which possess the property expressed by the formula $\lambda(\alpha)$. In other words, what is induced by a formula ... is not directly an existence, a presentation of multiplicity, but rather ... the "separation", within that presentation ... of a subset constituted from the terms ... which validate the formula" (BE 46). To write in abstract notation or indeed any language what a multiple is, you always introduce the separation between the collection as such (α), and the elements which are included in it $\lambda(\alpha)$. Every set is determined by the existence of the set as the collection of elements and the possible subset of those elements, and you can never speak of a set as such without also speaking of the {set} as well. What the axiom proves is the basic point we already made, that sets are collections of elements not fusions of things. This is true even if the {set} contains one element. If the {set} contains one element then you still differentiate between the set as container and the {set} itself of one element, meaning its cardinality is actually 2.

The axiom of separation states that every set has at least one subset so that we can always speak of every multiple as both a container and as something contained, or belonging and included. If we wish, we can take this axiom on trust and go straight to Badiou's explanation of the axiom as regards its ontological significance. This is, after all, the large part of what concerns Badiou as regards set theory and so also our own study. So, the importance for philosophy of the axiom of separation is that,

> the theory of the multiple, as general form of presentation, cannot presume that it is on the basis of its pure formal rule alone – well-constructed properties – that the existence of a multiple (a presentation) is inferred. Being must be already-there; some pure multiple, as multiple of multiples, must be presented in order for the rule to then separate some consistent multiplicity, itself presented subsequently by the gesture of the initial presentation (BE 47–8).

Concluding on the whole meditation here, Badiou is motivated to use separation to prove set theory in terms of realism rather than construction. What he says is that logic alone, the abstract notation $\lambda(\alpha)$, is not enough to present presentation, because the formula already admits to separation between the two terms. Rather, logic is what comes after a multiple is presented so that all forms of separation, sets of sets, subsets of subsets and so on, presume the existence of the multiple in the first instance even if that multiple as such is presented retroactively after the consistency of a situation of multiples as a set that has been constructed. This is clearly the basis for the retroactive logic of the final phrase, which describes the process of presentation of presentation, a tautological formulation we are already familiar with. What this implies is that for sets to be constructible using abstract formal language they first have to exist, as the axiom of separation shows that in order to describe a set as the elements which are included in that set, $\lambda(\alpha)$, the elements as such must already be presentable in said presentation. Multiples then, in order to exist, must be constructibly consistent. For them to be consistent they must submit to the most basic axiom of set theory, that of separation between set and subset. If they submit to this axiom however, then it is retroactively provable that multiples, in order to be constructible, must first be real. This is the basis of Badiou's material realism, a position of deep significance for him, especially when we consider in the second volume the nature of worlds. We could end now by saying the point of the axiom of separation is to prove that multiples are real, and move on. However, three elements of the argument leading up to this point are worthy of our extended attention. These are: the role of abstract notation, the self-predication paradox and the definition of the pure multiple as operative not existential.

Notation and self-predication

Badiou makes the point that Cantor's innovation of set theory is exemplary in the manner in which any conceptual event is always, in the first instance, totally inconsistent with itself. Cantor famously defines sets as regards grouping a totality into distinct objects due to mental intuition. Badiou notes that in fact, the effect of set theory is to render inconsistent all the terms of this operation: totality, objects, distinction and intuition. So that when Frege and Russell turned to set theory after Cantor, they refused to define a set in his terms, as regards mental intuition of objects, and instead moved away from mathematics relying entirely on formalized language to define the essence of a set[9]: first order predicative logic to be precise. In first order predicative logic a set is simply what validates the formal language of first order predicative logic. 'In other words, "set" is what counts-as-one a formula's multiple of validation' (BE 39). This seems very close to our previous consideration of indifferent, retroactive axiomatization as regards the presentation of the presentation that is the multiple. A set is what comes to be defined consistently as a set using the language of set theory based on a limited number of operations totally transmissible across a substantive community of mathematicians and logicians.

The problem with this solution of set constructability, correct in all parts as regards moving away from sets as collections based on intuition of objects, is that it is overburdened with paradox. The most famous of these is Russell's paradox expressed in his letter to Frege in 1902:

> Let *w* be the predicate of being a predicate which cannot be predicated of itself. Can *w* be predicated of itself? From either answer follows its contradictory. We must therefore conclude that *w* is not a predicate. Likewise, there is no class (as a whole) of those classes which, as wholes, are not members of themselves. From this I conclude that under certain circumstances a definable set does not form a whole.[10]

Although self-predication initially seemed to undermine Frege's logicist project alone, what it went on to show, by virtue of Gödel's two incompleteness theorems, was that no system can demonstrate its consistency from within the axioms of that system alone. Both Russell's paradox and Gödel's incompleteness theorems concerned systems based on statements about themselves or, if you will, self-mentioning constructible systems of the kind Derrida famously attacks in his paradox of margins and centres in his breakthrough essay 'Structure, Sign, and Play in the Discourse of Human Science'.[11] The self-predication paradox is *the* paradox of philosophy since its inception, in that no philosophical system has been able to found itself in a manner which is consistent with its own axioms if those axioms are founded within the system itself. This is also true of the philosophy of mathematics as Gödel showed definitively.[12] In light of this, the large part of Meditation Three is a consideration of Russell's paradox or, it cannot be the case that something can belong to itself if it is defined as the set of things which do not belong to themselves, taking belonging to mean as regards a collection, not a fusion. This being the case, the obverse is also true, no self-founding set can found itself from axioms founded within it. So just as a set cannot be consistent if it includes within itself the fact that it is the set of what it doesn't include, so a system for sets {set theory} cannot be consistent if it is a system that is based on axioms that it *does* include. By my reading, at least, Russell's paradox has this double-faced nature in terms of the paradox of inclusion of non-inclusion *and* inclusion of the law of inclusion as such, locating paradox in naïve, non-axiomatic set theory such as that of Cantor, at both the lowest level and the highest.

Although I have said that this is central to modern thought in terms of analytical and continental philosophy, we dwell on it here specifically because it relates to indifference as a 'solution' to the impact of self-predication on the definition of the multiple. What Badiou realizes is that the indifferent abstraction of set theory notation solves the related problem of what Kant calls philosophical indifferentism, or the inability to choose one system over another from within the consistencies of those systems.[13] One might also speak here of Derrida's point that every self-consistent system requires an inconsistent element to found said consistency in such a fashion as it undermines its own consistency.[14] Or Heidegger's point that in the formula A = A, the first A is the same as the second A, but not equal to it as the distance required for A to 'see' or better present itself as A, means A is never entirely given in the notation of A. Heidegger calls this 'heeding the difference as difference',[15] Gödel calls it incompleteness, Russell defines it as paradox, while ZF set theory, Potter and Badiou call it separation. They are all speaking of the same issue in varying forms of notation, or various languages. Finally, it is the indifferent nature of sets that allows Zermelo and then Fraenkel, von Neumann and Gödel to come up with the axiom of separation, which renders irrelevant the self-predication paradox as regards the nature of a multiple, stressed by Russell in terms of paradox and the need for his theory of types to solve the problem of set self-predication.

If, as Cantor, Frege and Russell do, you take the set to be a thing which can or cannot be defined due to your own system's formal consistency, then you will end up with the disappointing paradoxes I have detailed. Like Badiou, however, I am unimpressed by the apparent revelatory nature of Russell's paradox in relation to sets. Instead, all the idea of a set shows, is what you need to let go in your conception of the set in order to avoid self-predication, and Badiou says this consists of two things. First, you cannot define what a set 'is' as a thing because 'it is of the very essence of set theory to only possess an implicit mastery of its "objects" (multiplicities, sets): these multiplicities are deployed in an axiom-system in which the property 'to be a set' does not figure' (BE 43). Set theory is about using sets not defining them as objects with predicates. The existence of sets is proven by their operativity not their object-predicate relationality.

Second, any future set theory must avoid any inconsistencies and contradictions. To do this is fairly simple, it transpires. The presupposition of the unity of contraries in the oppositional formula being-of-the-one, a formula we began by showing was, is and will ever be a logical impossibility, needs to be abandoned. If we take being instead as not-one or as-not, and the multiple as pure presentation of being due to indifference, then we will never succumb to the paradox of self-predication. As Badiou puts it: 'Axiomatization is required such that the multiple, left to the implicitness of its counting rule, be delivered *without concept*, that is, *without implying the being-of-the-one*' (BE 43). If we take a multiple to be a pure, indifferent presentation of being as-not, which is what is meant here as being present without a concept, then quite simply the self-predication paradox is impossible, because it depends on a differential conceptualization of a multiple as a predicate of a concept, which is not what a multiple is according to set theory. What we can observe here is that indifference as concept-neutrality solves the paradox of differential identity intrinsic to Russell's formula. This allows us to say that a set is nothing that exists as an object of a class. It is a purely abstract, concept-neutral, indifferent operation. As such it cannot be captured by first order predicative logic or subjected to the paradox of self-predication because a set is not a thing possessive of predicates.

The pure multiple is real

The rejection of Russell's paradox and the liberation of multiples from the confines of purely constructible systems such as first order predicative logic, lead us to our third and final point, the definition of a pure multiple. This is also the definition of a pure set, origin of the paradoxes we have just considered, because multiples are in fact sets or more accurately, all sets are reducible to pure multiples. And so we come to the first and only true presupposition of set theory, the one axiom we need to take on trust, from which all others can be induced or deducted, the axiom of belonging. As Badiou says, the 'entire lexicon' of mathematics is reducible to belonging. This being the case it is crucial that we stipulate precisely what Badiou's sense of belonging determines, not least because we cannot rely on an intuition of objects (Cantor) or pure abstract notation (Frege, Russell) when it comes to its definition. Badiou says, in a masterful statement:

> The multiple is implicitly designated here in the form of a logic of belonging, that is, in a mode in which the 'something = α' in general is presented according to a multiplicity β. This will be inscribed as $\alpha \in \beta$, α is an element of β. What is counted as *one* is not the concept of the multiple; there is no inscribable thought of what *one*-multiple is. The one is assigned to the sign $\in$ alone; that is to the operator of denotation for the relation between the 'something' in general and the multiple. The sign $\in$, *unbeing* of any one, determines, in a uniform manner, the presentation of 'something' as indexed to the multiple (BE 44).

We can deduct from this that, first we can see that indifference, something = α, is always 'relational'. Second, that yes, one can speak of something belonging to something else in a purely presentational, indifferential, quality-neutral fashion. This is possible because the one is-not and so all the paradoxes of the one-multiple right up to and including Russell's, are rendered irrelevant to the task. One is not something, either an object to be collected or a collection of objects; rather one is the operational possibility of belonging as such. Belonging operationally relates the indifference of something to the indifference of some things if you will. Belonging, on this reading, has a primary role in negating the presupposed being of the One, and when it does so, separation between presentation of presentation and presentation itself, the two aspects of the multiple, is possible without the ancient and modern aporias attendant on that. In our system, we would say that belonging is an indifferent relationality between something in general, absolute in-difference as devoid of quality, and the indifference of the multiple, pure difference as such devoid of quality.

A dramatic *epoche* or reduction now occurs. For example, α and β are entirely substitutable in the notation and can be substituted for 'specifically indistinguishable terms' as all that separates them is the operation of belonging. Not only does this reinforce again our point about indifference, here as a form of indistinguishability, but it also explains how relationality can be taken to be nonrelational. If what relates α and β is the symbol $\in$ as an operation of unbeing as regards the one, then belonging, the primitive base of all relation which is, after all, the pure definition of logic and mathematics, is-not and therefore, by definition, is nonrelational. Belonging is the application of the nonrelational to an operation of relation so that relation can occur without paradox or incoherence. How do we know this? Because if we allow this to be true, set theory works perfectly to such a degree as now there is little or no debate about the possibility of mainly reducing mathematics to set theory. Not as the foundation for mathematics as an object, but for mathematics as a world-wide, never-ceasing and entirely consistent and transmissible daily activity.

We are then led by this fact to a second reduction due to indifference. There are no sets of multiples or collections of elements. Sets and multiples are indifferentiatable: 'all is multiple, everything is a set' (BE 44). Badiou provides a proof of this but for our purposes it is better to dwell less on the proof and more on its implications. For the record, the proofs of the various axioms of set theory do not come from Badiou, but, importantly, in each case their philosophical significance does. For example, he asks us to accept that: 'what belongs to a multiple is always a multiple; and that being an "element" is not a status of being, an intrinsic quality, but the simple relation, to-be-element-of, through which a multiplicity can be presented by another multiplicity' (BE 44–5). If this is so then set theory shows that it never *speaks* of the *one*, but instead *presents* the *multiple*. Or, 'Any multiple is intrinsically multiple of multiples: this is what set theory deploys . . . a property only determines a multiple under the supposition that there is already a presented multiple' (BE 45).

If you prefer, you can think of this as the basic rule of all systems of immanence. One can find it in Foucault, Deleuze and many other places. It is also intrinsic in some senses to the constructivist positions or all formal, that is non-realist, modes of analytical thought that Badiou is moving away from here by his powerful refutation of the construction of sets as objects in first order predicative logic. This is why the final comment on pure multiplicity is important: 'Zermelo's axiom system subordinates the induction of a multiple by language to the existence, prior to that induction, of an initial multiple' (BE 45). This is what the axiom of separation that we began with proves. All notation, all language, depends on a primitive sense of belonging or multiplicity to function, therefore, if multiplicity precedes the notation $\in$, Plato is right, something 'real' exists prior to our linguistic denotation of it.

After proving this more formally in the pages which follow, Badiou is confident enough to say that language always infers a sub-multiple, so that 'Language cannot induce existence, solely a split within existence' (BE 47). The axiom of separation establishes 'that it is solely within the presupposition of existence that language operates – separates – and that what it thereby induces in terms of consistent multiplicity is supported in its being, in an anticipatory manner, by a presentation which is already there. The existence-multiple anticipates what language retroactively separates out from it as implied existence-multiple' (BE 47).

Accepting the axiom of separation and its wider implications for the material reality of the multiple as the basis for the linguistic expression of its constructible consistency, Badiou's main aims in this meditation, are all discipline-changing propositions. But Badiou closes out the third meditation in admitting they are only supportable and potentially transmissible (communicable) if we can indicate the location of the initial starting point of being and, for that matter, how you stop the multiple of multiple structure proliferating into bad infinity. We have come so far and yet we are still mired, it would appear, in the ancient swamp of Aristotelian classes: infinite regress and infinite proliferation (substances and gods). It is time to look directly at being as-not in terms of its being what Badiou calls the void, so as to establish definitively that ontology can avoid the paradoxes and logical impasses of Western thought if it defines being as the reciprocal relation not of the one and the many, but of the void and the multiple.

The void: proper name of being (Meditation Four)

The fourth meditation concerns Badiou's choice to name being, which is-not, the void. Three aspects of this choice are debated. That being and the void are not the same, as the void names being. That the void and nothing are not the same, as the void names a local subtraction of being from a specific situation, while nothing names the global necessity to subtract being to have any presented situations at all. Finally, that while void 'names' being, it does so through an act of pure nomination: the proper name of being. Although these stipulations are very specific, they are balanced by the fact that the entirety of ontology is composed of one 'term', the void. This then is a fourth element up for consideration not least because this assertion is based on the indifference of void, an entirely empty term, to any other possible, conceptually full terms. One final point ought to be made. As we have already stated, the fact that being is-not and that local to a situation it can be called void, does not mean that Badiou is engaging in any kind of negative onto-theology, committing to being in withdrawal through the back door or presenting a metaphysics of alterity of any order.[16] When we say being is-not globally, we mean there is no consistent whole, and when we say it is void locally, we are speaking of its very tangible operational effects in a local situation which, however, due to the nature of its operational axioms, it cannot be present to or existent in.

The meditation begins with a reiteration and summary of the basic ontological position we keep revisiting here: How can it be that the one is-not? In this context Badiou reconfirms this axiom as regards the nature of the count. 'Insofar as the one is a result, by necessity "something" of the multiple does not absolutely coincide with the result' (BE 53). This remainder or phantom 'cannot in any way be presented itself' because to be presented means to be counted, while clearly here we are dealing with what precedes the count so that the count can be a result. This leaves us with only one option as regards the relation of the precedent of the count-as-one, a subtracted being: 'one has to allow that inside the situation the pure or inconsistent multiple is both excluded from everything, and thus from presentation itself, and included, in the name of what "would be" the presentation itself, the presentation "in-itself"' (BE 53). From our perspective, if we take pure presentation as such to be indifferent, and again here the subtraction from the situation is defined as an indifferent 'something', then what Badiou is moving towards is naming this something. He does not name it indifference, in fact, but rather the void. That said, the void is the first, clearly indifferent element of his ontology. Furthermore, in keeping with Badiou's own concentration on operations indifference is, like belonging or being as subtraction, not a thing or a concept but a process.

In speaking of precisely how the void can be purely subtractive to a situation of presented multiples, Badiou makes the point that the count, structure, must be absolutely watertight. Structure, what he later calls nature, is completely stable. Although founded on inconsistency, and always inclusive of inconsistency, inconsistency is-not and so consistency is totally consistent. This is what allows Badiou to speak of the universalism of consistent situations. One can see, I hope, the ontological need for this and the logical proof of it. Further, if nonrelationality is accepted as regards being, then this is also absolutely complete. There are no half-way houses for nonrelation, floating indifferent monads ripe for exploitation representing what Deleuze calls the indifference of pure determination. Badiou says as much when he explains in regard of structure 'the law does not encounter singular islands in presentation which obstruct its passage. In an indeterminate situation there is no rebel or subtractive presentation of the pure one upon which the empire of the one is exercised' (BE 54).[17] Just as being can never encounter a one, a count, so counting does not encounter rebel 'pure ones' as islands of nothingness, which undermine it. As we do not yet have to contend with the idea of the event we can say that the count, a structure, indeed all consistency, in that it is operationally dependant on being as that from which counting is the result, means that ontology is absolutely stable. The one thing that can undermine consistency at this stage, inconsistency, is the thing that consistency is founded on as that which is subtracted from it.

The void and nothing

Having named being as void, Badiou turns his attention to some traditions of the presentation of nothing, poets in particular, and the dangers of taking nothing as a kind of a thing, a nothing, which up to this point we have been doing to better clarify the positive presence of the void in relation to the count. Moving on from this early position he gives a clearer sense here of how he 'perceives' nothing: '"nothing" is what names the unperceivable gap, cancelled and then renewed, between presentation as structure and presentation as structured-presentation, between the one as result and the one as operation, between presented consistency and inconsistency as what-will-have-been-presented' (BE 54). Yet, in contradistinction to the uses made by most philosophers of difference, in Badiou's ontology this inconsistency is resolutely nonradical because 'nothing' is intrinsic to the count and thus totally banal. Difference, inconsistency, is always included in structures of presence, identity, consistency and so on. This being so, on its own, inconsistency taken as difference has no potential to disrupt structures.

Badiou now gives his reasons for replacing nothing with the term void. We won't linger here except to say that void is preferable because it is local to a situation, while 'nothing' names the global effect of structure: '*everything* is counted' (BE 56). As it is essential that Badiou's ontology is immanent, naturally the void is the preferred term. It is not choice of term so much that concerns us, in fact, as the nature of void as a term. To discuss this, Badiou prefaces the consideration with a summary of the three key concepts thus far. First, ontology is a presentation of presentation or theory of multiple of multiples without a one. Second, that its structure is determined by an implicit count available through axiomatic retroaction without a concept-one. And now third, the void is the sole term of ontology. Let us now take it as read that ontology is presented due to the term void, we have done enough work on this at this point, and instead concentrate on the other contention here, that void is the sole term of ontology. The two are related as Badiou explains:

> if one supposed that ontology axiomatically presented other terms than the void . . . this would mean that it distinguished between the void and other terms, and that its structure thus authorized the count-as-one of the void as such, according to its specific difference to 'full' terms. It is obvious that this would be impossible, since, as soon as it was counted as one in its difference to the one-full, the void would be filled with this alterity (BE 57).

This, I would argue, is what has happened repeatedly in the philosophies of difference since Hegel. The only solution to this tendency towards the reification of alterity is that all terms are void, and this is due to the indifference of the void as term. If indifference were a term amongst terms, it would be necessary to differentiate it on at least two counts. First, it is one term amongst several others and so differentiatable from them due to content, but it has no content. Second, if it were a term that one could differentiate from all other kinds of terms, the empty term as different from the full, as Badiou says emptiness would fill the void as a quality. So the void cannot be an identity, as it cannot be counted, nor can it be taken as any order of difference, because it would take difference or 'emptiness' as a quality. The only solution therefore is that the void is in-different both in terms of being entirely devoid of content, and in terms of it being impossible to differentiate from other terms. One can say if you wish, after Badiou, that because the void is the only term it is necessarily indifferent, or after my own reasoning, because the void is indifferent it is the only term.

All of this leads to the question we raised in the previous section. If every multiple is a multiple of multiples, solving the ancient impasse of the one and the many, then what is the first multiple, as infinite regress was always the one shoal on which the bark of philosophy foundered? We now have our answer: 'the "first" presented multiplicity without concept has to be a multiple of nothing, because if it was a multiple of something, that something would then be in the position of the one' (BE 57–8). We have already covered this in fact by differentiating aggregation as fusion and as collection, set theory being the latter. As Potter showed, if a set is a fusion then an empty set makes no sense. But if a set is a collection then a collection with no members makes absolute sense. We will go on to see the mathematical proof of this via the axiom of foundation but all we need know at this stage is that a set can be empty. Of course it can, because a set is a multiple and a multiple is merely a presentation due to the void. More than this, nothing in the multiple makes it a multiple. All that makes it a multiple is that it can be counted as one. And all that facilitates the count-as-one is the inexistent 'presence' of the void in every count. Badiou adds that this solves a major problem. If being is presented as pure multiple, that which can be presented, then it is neither one nor multiple as it is presented through the void multiple. As the void multiple presents the void it is not one, because the void is-not, and it is not a multiple either, because it is empty. Again, my version of this would be that the void is indifferent, being neither a one nor a many, an identity nor a difference. 'This way ontology states that presentation is certainly multiple, but that the being of presentation, the that which is presented, being void, is subtracted from the one/multiple dialectic' (BE 58).

Void as nomination

The last thing to consider is the nature of nomination: being named as the void. So, to clarify, being is not the void. The void merely names/presents being. Being precedes all language, a key difference between Badiou's use of formalization and that of the majority of analytical thinkers. The void is not just nothing but nothing specific to a situation. It is local. Void is the only term but it is not a normal term as it names a process not a thing. This first of all allows us to explain how a void can be called a multiple when it is neither one nor many. Badiou says it is named 'void' so that it fits into the 'intra-situational opposition of the one and the multiple' (BE 59). Badiou says his hands are tied in this regard as for something to be presented it must be named. You cannot name the void a one as it is-not, and the only other element in presentation which is, you recall, a dramatically reduced universe depending entirely on the belonging of the one and the multiple, is to call it a multiple.

All of this means the name of the void is of a particular order. Naming here is not referential but purely ostensive and denotational. Badiou says 'because the void is indiscernible as a term … its inaugural appearance is a pure act of nomination' (BE 59). This means void does not name a thing, subject-predicate, but is a pure act. It does not say what something is but that something can now be, in accordance with our opening comments on the importance of the discursive law of communicability to Badiou's overall philosophy. Void then is a sanction of presence due to its use as a mode of mention, a formulation outside the bounds of analytical philosophy's limitations on language as reference. Badiou goes on: 'The name cannot indicate the void is this or that. The act of nomination, being a-specific, consumes itself, indicating nothing other than the unpresentable as such … The consequence is that the name of the void is a pure *proper name*, which indicates itself, which does not bestow any index of difference within what it refers to …' (BE 59).

The void, as name, is a pure, ostensive act of deixis. It is a content-neutral activity of pointing that is self-predicative. That you can name the void means that the void can be named. Linguistically, this is another version of the presentation of presentation only in reverse for, as Badiou insists, what the void presents is the unpresentable as such. In fact, the two are elements of the same moment.[18] The pure presentation of presentation depends on the unpresentable to come to presentation and then it allows this unpresentable element to be retroactively named as the void upon which presentation depends. The whole process is dependent on the basic law of ontology as Badiou expresses it: 'being, submitted to the law of situations, *must* present' (BE 58) or, if you wish to retain ontology you can only do so by accepting that being is-not and to do that you need to make sure being is always immanent to a situation to avoid the usual aporias. This means therefore that being must be presented and the only way to do that is to name it as a void multiple.

ZF+C: the nine axioms of contemporary set theory (Meditation Five)

It is in Meditation Five, 'The Mark', that Badiou begins to systematically address the fundamental axioms of set theory. And it is in the same meditation that he starts to freely use the term 'indifference' to describe the axiomatic processes of set theory. Badiou uses the totally orthodox Zermelo-Fraenkel theory of sets plus the axiom of choice (ZF+C). Ernst Zermelo proposed the first axiomatic set theory in 1919, a theory whose failings were pointed out by Abraham Fraenkel in 1921, and then rectified by Fraenkel and Thoralf Skolem in 1922. It came to be, for decades, the dominant mode of set theory mathematics with even the initially controversial axiom of choice being widely accepted. ZF+C presents a total of nine axioms. Of these nine, Badiou needs only the first six to establish his ontology, plus one primitive existential: belonging. A primitive existential is an element of the system that cannot be proved based on other axioms. Consistent systems like to reduce primitives to a minimum. Set theory is exemplary in having only one, belonging, and in the immense number of consistent operations this facilitates, meaning that while you cannot 'prove' belonging, the only way you could question its operative veracity is by never using set theory and totally rejecting the idea of Being rendering noncommunicable both mathematics and philosophy for the last hundred years.

Of these six axioms the sixth, axiom of the void set or there exists a set which has no element, is the prize for Badiou. The other five are placed in service of the sixth because it is the sixth upon which depends one of Badiou's essential concepts: subtractive being. I will only detail these axioms if necessary to our wider project of indifference or if Badiou's own comments throw penetrative light onto the essential detail of his own ontology. For now, let us just look at how the first five axioms define the basic universe of ontology. These axioms are extensionality, power set, union set, separation and replacement. Without yet going into detail Badiou says that the axiom of extensionality 'fixes the regime of the same and the other' (BE 65). The power and union sets are to do with the internal organization of sets. Power sets concern subsets and so 'regulate internal compositions', while union sets concern dissemination within sets. The crucial stipulation here is that within sets the entirety of structure is 'under the law of the count' (BE 65–6), so that nothing is encountered in a set that is lower down or higher up than the set. As this is the essence of *Logics of Worlds* we don't need to overthink structure here except to say the point is that these two set axioms, power and union, maintain the internal consistency of every presented situation.

If extensionality deals with same and other, power and union sets then distribute identity and difference internal to a set. These three axioms effectively treat most, if not all, of the aporias of the consistency of Being thought of in relation to identity and difference.

We have already considered the axiom of separation because this was so necessary for the subordination of language to the already-being of presentation as such, Badiou's realism through a negation of the linguistic turn in philosophy of the last hundred years. So, to be neat, the first three axioms revive metaphysics and negate twentieth century philosophy of difference. The fourth negates the 'linguisticism' of most analytical philosophy. The fifth axiom, of replacement, vouchsafes the content-neutrality of elements of a set, making them absolutely replaceable one by one. It is one of the central axioms of indifference in sets but here for Badiou it simply helps us remember the purely indifferent nature of the multiple. Of these five, we will linger on the first, extensionality, and the fifth, replacement, neglecting the other three for the present for the reasons just given, so as to arrive at the sixth: the void set.

Axiom of extensionality

Badiou begins by laying out three elements of set theory in relation to ontology. First, that we accept ZF plus choice or nine axioms. Second, that we never try to prove the existence of the multiple (axiom of separation). Third, we accept one primitive, belonging. Now we have to engage with axiomatic extensionality as Badiou's explanation of extensionality shows that it is just another way of formally resolving the millennia-old paradoxes of self and other, one and many, transcendence and immanence, and so on. 'The axiom of extensionality posits that two sets are equal (identical) if the multiples of which they are multiples, the multiples whose set-theoretical count as one they ensure, are "the same"' (BE 60). The axiom of extensionality is another way of stating Hegel's conception of indifferent difference of $A \neq B$ without assigning any properties to A and B. It allows one basically, in an operational count, to designate identity and difference in a totally content-neutral and thus purely indifferent fashion.[19] To put it simply, two sets are 'the same' if they count 'the same', if they can be counted in terms of indifferent identity. And naturally if this is possible then the obverse is true as well, two sets can be counted as different based on the same indifferent law. This is why the indifference of pure difference as such can be called indifference or in-identity because the one thing that metaphysics teaches us, incorrectly as it transpires, is the foundation of being on the co-dependency of identity and difference.

Badiou explains this as follow. The axiom of extensionality 'indicates that if, for every multiple *y*, it is equivalent and thus indifferent to affirm that it belongs to α or to affirm that it belongs to β, then α and β are indistinguishable . . . The *identity* of multiples is founded on the *indifference* of belonging' (BE 61). That is, if we accept the multiple to be indifferent, meaning we do not try to define the multiple as a quality possessive thing, then we also accept that multiples and sets are indistinguishable: all sets are multiples, all multiples are sets. This is a neat solution as you can found the same on the same, one of the vicious circles of metaphysics, because you have modified what 'the same' means. If you say two sets are identical if they have the same elements, all you are saying, Badiou reminds us, is that 'element' here is not anything intrinsic but merely the result of the count or one multiple as presented by the presentation of another. As this is such an important part of indifference and Badiou's ontology I will reproduce the formal notation: $(\forall\gamma)\ [(\gamma \in \alpha) \leftrightarrow (\gamma \in \beta)] \rightarrow (\alpha = \beta)$. What the formula shows is that for all *y*, if *y* belongs to α and also belongs to β then by definition α is the same as β. If this marks what we mean by the same, then it also, perhaps more importantly, determines what is different. So, if *y* 'maintains a relation of belonging with α – being one of the multiples from which α is composed – and does not maintain such a relation with β' (BE 61), then the two sets are counted as 'different'. This is the precondition of all difference as such, pure indifferent difference.

And if the reader is in any doubt, aside from considering Badiou's own comments on indifferent difference in Hegel, they can also consult the following passage on difference due to extensionality: 'What possible source could there be for the existence of difference, if not that of a multiple lacking from a multiple? No particular quality can be of use to mark difference here, not even that the one can be distinguished from the multiple, because the one is not' (BE 61). If we take a being or a stable identity as what can be counted as some thing, which set theory depends on using formal abstract notation determining a thing entirely by the specificity of its relations, then defining some other thing as different can only be that it is not included in that count. If we take otherness to be something extrinsic to the count, then we are determining it as some thing countable as not being counted. But, as we have shown, such a position is disallowed due to an endless number of logical impasses. If, however, we accept that whatever is not counted is-not, then all we have to say is that difference is a multiple that is not counted. It is not a thing, actually existing, that we forgot or neglected to count, it is simply the operational definition of difference due to that of identity or, due to the count we can found the same on the same without a viscous circle, and we can found difference on the same without counting the differential element as some thing. Badiou ends with a warning that this axiom does not tell us if anything exists, all it does is say if there is such-and-such a multiple, then we can also determine what is not such-and-such a multiple. If there is the same, then we can say that there is the different. Thus we need the axiom of separation, multiples exist, before we can sign up to the axiom of extension.

Axiom of replacement or substitution

The axiom of replacement or substitution relates to, and is the natural concomitant of, that of extension. If a set is indifferently defined entirely by the operativity of the count due to the movement of relation between multiples as elements and multiples as sets, then if you replace every element of a set with other elements, one by one, you will end up with the same set. This is the axiom of replacement. Naturally, like the axiom of extension, replacement is counter-intuitive to any idea of a set as a collection of things due to qualities they share. Rather, in set theory 'if a multiple of multiples is presented, another multiple is also presented which is composed from the substitution, one by one, of new multiples for the multiples presented by the first multiple' (BE 64–5). Commentators do not spend much time on this axiom, yet for us it is remarkable again for the way Badiou goes on to describe its significance. He himself notes that the idea is 'singular, profound' (BE 65) as it is

> the equivalent to saying that *the consistency of a multiple does not depend upon the particular multiples whose multiple it is*. Change the multiples and the one-consistency – which is a result – remains ... What set theory affirms here ... is that the count-as-one of multiple is indifferent to *what* these multiples are multiples of ... the attribute 'to-be-*a*-multiple' transcends the particular (BE 65).

What the axiom of substitution shows is if we summarize what it means to 'be' something, which means to be a multiple, the definition of one's being has no particularity. The one thing that seems most obvious about being, that you are an object with particular predicates that you share in common with other objects, and so you share identity, and do not share with others, and so you confirm identity through difference, is here entirely negated by two elements: the one as a result of indifferent counting. Identity has nothing to do with predicates.

Difference has nothing to do with qualities. Everything is entirely substitutable with everything else provided it is done step by step, and that in the end the 'number' or the set, its cardinality, is the same. This is, after all, all the count is, the combination of one by one and an overall result. Badiou ends his rather brief consideration of replacement by simply confirming something that I hope is apparent, which is that the law of pure presentation as such is entirely dependent on the axiom of substitution.

The void set and in-difference

Badiou's subtractive ontology is due to the sixth axiom of the void. The void set, we have already explained, is the proper name given to being which is-not in a local situation. It is necessary to allow being to present itself as inexistent to a count in a situation. It is also necessary as the minimum limit of a count to avoid infinite regress of sub-division. Finally, it can be said to be in-different as we saw. Badiou begins his consideration of the axiom of the void set, rehearsing most of these issues and of course the danger of taking the void as some thing, and hence countable. To avoid all the aporias of ontology and nothingness, which are considerable, there is one pre-requisite for the void 'it cannot propose *anything* in particular' – as we have seen, presumption of object due to possession of predicate is where ontology always collapses into paradox. In contrast, holding to an indifference of particularity due to object existence by virtue of quality, means we can also have an indifference of difference:

> the difference between two multiples . . . can only be marked by those multiples that actually belong to the two multiples being differentiated. A multiple-of-nothing thus has no conceivable differentiating mark. The unpresentable is inextensible and therefore in-different. The result is that the inscription of this in-different will be necessarily negative because no possibility – no multiple – can indicate that it is on *its* basis that existence is affirmed (BE 67).

Ostensibly, Badiou is explaining the necessity for the void to be a proper name of being, a name which inscribes negation. Yet what interests us is this use of the term in-different. We have already explained why the void is indifferent due to it being quality neutral thanks to its relation to pure presentation as such which is indifferent. Now, we can add that the void names the in-different of a different order, that which cannot ever be differentiated. If relational differentiation depends on the axiom of extension, then the void, which is not presented in a situation is, by definition, without differentiation. Yet at the same time because it is-not, it does not have an identity either so it is in-different. Here indifference is not a quality of the identity-difference copula, but something 'apart' from that copula. Due to the void it is possible to speak of some thing, albeit in the negative as pure operational result and based on the fact that it is-not, which is impossible to differentiate and which yet is not pure immanence either. 'The unpresentable is that to which nothing, no multiple, belongs; consequently, it cannot present itself in its difference' (BE 67). This is what we mean by the term in-difference.

Badiou goes on to explain this in formal, mathematical terms which we do not need. That said, before we leave the void as proper name and indeed the first section of *Being and Event*, there is one final stipulation as regards the void that cannot pass without comment. After having explained the void set as presentation-multiple-name, and the formal notation for the axiom 'there exists β such that there does not exist any α which belongs to it' (BE 68), Badiou turns to the problem of one versus unicity. He says that one and unicity are not the same but this does not mean something counted as one cannot also be unique, as all it shows is that this multiple is different from any other. Yet if the proper name as pure proper name, which is what the void is, is to function, it must be unique. It has to be an operative, deictic ostension that names something through the unique operation of its naming. 'However,' he realizes, 'the null-set is inextensible, in-different. How can I even think its unicity when nothing belongs to it that would serve as a mark of its difference?' (BE 68). This raises technical problems such as the possibility that you could have several voids by thinking of it as a common name, a species, with all the age-old problems of class that we began with. What makes the void unique is what disallows one from ever thinking of it as a something at all so you cannot think of being-void as 'a property, a species, or a common name ... The truth is rather this: the unicity of the void-set is immediate because nothing differentiates it, not because its difference can be attested. An irremediable unicity based on in-difference is herein substituted for unicity based on difference' (BE 68).

You cannot think of the void as 'different' in any way that has been available thus far to predicative thinking since the Greeks. The void's in-difference is the result of not being available to be differentiated. It is something that precedes and founds extension, something in this sense real and not due to extensional language games (logic).[20] This is what makes it unique, because it cannot be differentiated, but it is a uniqueness that is operational due to the process of being able to name it. Each time you name the void as name you do not name the 'same' thing, as it has no qualities and it is-not. Nor do you name something unique and singular each time. Rather, you name the unique nonrelationality of being as void as a process due to its subtractive inexistence. By definition, the void is unique not because it is something you cannot establish a relation with, but because it is what allows you to establish relation due to the operational presence of absolute nonrelationality. In short, being is nonrelationality as in-difference.[21]

Conclusion: pure multiple and the void

We learnt that in terms of sets being indifferent collections rather than predicative fusions, set theory mathematics presents us with a primitive definition of sets that appears, at times, counter-intuitive. Indeed, sets are determined by one simple primitive that precedes logical relationality: the clear division between there being such-and-such a set due to a determinate number of members, and what can be collected within that collection. From this point on then we need to accept that the key defining aspects of sets is that they are indifferent and bifurcated, qualities which will facilitate the shift from talking of sets to speaking always in terms of multiples of multiples. We then considered that the axiom of separation provides, in set theory, a formal proof for the fact that sets are indifferent collections not predicative fusions.

One crucial implication of separation is, contrary to the dominant tradition of taking sets of multiples as logically consistent constructions, the fact that you cannot but present a set as bifurcated, even if the set is empty, means that there must be something before the formula $\lambda(\alpha)$, in other words (α), in order for the formula to function as the basis of all sets. There is, in other words, before every single collection or set, a pure multiple as such which the set is the collection of, including if the set has one element only, and true even if the set is empty. If a set has only one element, it still has a subset, and if a set has no elements, it can still be called a collection. The pure multiple which precedes every set but which can only be designated retroactively after the formation of a set is what Badiou means, after Plato, that there is something real in being before beings in the world can be constructed. One important implication of Badiou's materialism is the relation of sets to the self-predication paradox. Badiou shows how the indifferent and bifurcated nature of sets as collections means that they do not succumb to the paradox of the set that belongs to itself. In that sets are not paradoxical in this way, Russell's theory of types and the many innovations in constructible worlds which follow are not necessary. This position implies that many constructible theories of the last hundred years or so misrepresent that basic nature of the multiple when collected into a set. Gödel, Heidegger, Dummett and Derrida all composed complex philosophical systems based on the presupposition that 'multiples' can only be constructed (they don't all use the term multiples). In contrast, Badiou believes that separation proves the opposite: that multiples are by definition real. That said, we also have a warning here. Badiou commits himself to the negation of sets as self-predicative only on the condition that any future theory of sets must avoid being defined as self-predicative. Just because sets are not self-predicative does not mean that self-predication is irrelevant. In fact, as we shall see, Badiou's commitment to avoiding self-predication will have dramatic effects across both volumes of the Being and Event project.

Badiou now replaces the logical, Cantorian formula of sets $\lambda(\alpha)$ with the set theory formula of multiples $\alpha \in \beta$, or α is an element of β. The importance of this change in notation is that it foregrounds the one, primitive element of set theory, belonging or $\in$. A second significance is that it no longer suggests that there is a difference between sets and their multiples. In fact, sets are indistinguishable from multiples in set theory and from this point on Badiou will primarily prefer to speak of multiples rather than sets because it is the ontological nature of multiples that bothers our traditions, not necessarily their collection into sub-set units. What this allows us to achieve is the realization that when we think we are speaking of the one what we are really talking about is the multiple which is, as we shall see, always a set or multiple of multiples. It also permits us to appreciate that the one in this context is the operational consistency of the count of otherwise inconsistent elements as belonging together such that they are included as elements of a single collection or aggregate. The final point to be drawn from this change in notation is that while $\lambda(\alpha)$ depends entirely on linguistic constructability, missing the crucial point of separation, $\alpha \in \beta$ does not. Instead it asks us to accept the designation $\in$ or primitive belonging on trust as non-constructible. That said, trust will be rewarded as retroactively, due to separation in the first instance but also many other operational proofs, the primitive, material, reality of $\in$ or of belonging will be demonstrated, a million times and more, to be a powerfully consistent position by the mathematical community.

We then moved on to the proper name of being. If being is-not globally, when it is presented as subtracted from a particular situation instead we call being the void of the situation. This helps us differentiate being from other problems surrounding the conception of nothing, and reminds us that irrespective of our axioms being is-not, being always inexists in a specific situation so we always encounter being as the named element of that which is subtracted from a situation such that the situation can present itself as consistent through representative counting. This subtracted element is called the void. Having explained the nature and necessity of the name void, we also showed how the void is the first really tangible proof that Badiou's philosophy is indifferent in nature. The void is in-different in terms of being entirely devoid of content, and in terms of it being impossible to differentiate from other terms, meaning that it is the only term that can be used to designate being. If there were more than one term we would be suggesting one term would be a more full presentation of being than the other suggesting being would be quantifiable across two or more terms. This would reduce these terms to predicates of the concept of being negating the very essence of being. The final conclusion to be drawn from the nature of naming pure being as it is subtracted from a specific situation the void, is that the void as name is an act of pure ostension. Naming being does not describe being, rather it simply points to being as inexistent to a specific situation so that said situation can be said to be totally consistent. Void does not name a thing as being, rather it indicates a process whereby being is subtracted from the count of situation so that the situation can be said to be counted.

Having established that Badiou uses the nine axioms of the set theory called ZF+C, we then concentrated on the axioms of extensionality and replacement as the main facilitators of the consistency of the sixth axiom of the void set which, as we already explained, is the foundation stone for Badiou's ontology due to realist implications of the axiom of separation. The axiom of extensionality is just a formal means of proving the basic tenet of indifferent difference such as it is presented in Hegel: $A \neq B$ taking A and B to be content neutral yet still differentiatable. Hegel struggles with the implications of indifferent difference as do all philosophers inheriting his solution to the problem: a philosophy of difference. The axiom of extensionality demonstrates that in fact indifferent difference is not a problem to be solved by difference but a solution to the aporias of a philosophy of identity and difference. Badiou then goes on to discuss how the axiom of substitution, all the elements of a set can be replaced one by one by other elements and it remain the same set because it has the same size or cardinal number, proves that multiples are indifferent or quality neutral. The importance of extension and substitution for our study cannot be overstated. Not only do they show Badiou openly admitting that his theory of being is indifferent, they also provide formal proof for two basic tenets of indifferential philosophy. Extension proves the consistency of indifferent difference, and substitution proves the content neutrality of every multiple. In other words, extension + substitution prove the tenet of indifferent difference due to the quality neutrality of beings in any situation.

Badiou finally returns to the axiom of the void set revealing, for us, a clear tripartite sense of the role of indifference in terms of his ontology. We were able to speak of indifference as indifferent difference, as quality indifference and now as absolute in-difference or total nonrelationality. We could say this in another way by stating that all situations are indifferently relational due to the quality neutrality of every element of the situation and the in-difference of the void set subtracted from every situation. This compound of three versions of indifference is what allows Badiou to resolve the longstanding aporia of singularity as the unique by showing that the void is able to name being as unique due its being in-different. In particular to solve what he believes to be the question of contemporary philosophy: 'what exactly is a universal singularity?' (TW 82).[22]

We are now at the end of the first part of *Being and Event*. We have proven that being is-not. We have defined the nature of the pure multiple. We have said that all that exists is multiples of multiples. We have defined sets in terms of collections such that we have proven that all sets are bifurcated multiples. This means we can both say that multiples are real, and that the void set exists such that we can name the local being subtracted from a situation as the void. Finally, we have shown how the application of set theory to prove these points depends fundamentally on the different but relational sense by which we can say collections are indifferent relations, elements are quality neutral ones and the void is in-different being subtracted from every situation. Our work on the indifference of pure being is, in a sense, over, and Badiou moves forward to consider not so much the void of presented being as the excessive nature of being when it comes to the nature of sub-sets of multiples due to the means by which belonging founds inclusion. Our job then is to say, having proven that pure being as void is indifferent, that this indifference is carried over in a significant fashion when we stop speaking of the pure multiple as such, and consider the multiplicity of multiples that makes up every represented situation.

3

Being and Excess

While sets and multiples appear to be two different entities, they are in fact different ways of expressing the same thing. When a collection of multiples is counted as a collection it is called a set. Yet every set is also a multiple of another multiple. Consider this multiple a set when it counts downwards, if you will, and a multiple when it counts upwards, so to speak. The obverse is true. A multiple is always a multiple of a multiple but, by the same logic this means it is a multiple composed of other multiples and so it also always a set. This double-vision of multiple-set is just another way of stating the one-many reciprocal relation of Western thought, meaning that at some stage the multiple will have to submit to the law of infinite subdivisibility to answer the question: If every multiple is a multiple of multiples what is the first multiple? And of course the problem of bad infinity: If every set is a multiple of another larger set, what is the largest set of all? Set theory then locates us in the same territory as metaphysics but, what we shall see is, in using multiples and sets, it is able to formally prove beyond any reasonable doubt that infinite regress and bad infinity are no longer inevitable when it comes to the nature of the being of all beings.

The relation between multiples and sets is determined by that between belonging and inclusion in set theory and this is what we are now going to concentrate on. Not what constitutes the pure multiple but how multiples are combined to form other multiples. How this combination includes a point of excess which actually determines the consistency of every set. And finally how, for multiples to be composed into sets, first all the elements of a set must be counted or must be said to belong, then they must be, according to Badiou at least, re-counted, so that they can be said to be included as subsets of the consistent set. What the second part of *Being and Event* is concerned with then is the nature of sets, the different ways of counting sets, in terms of what belongs to them and what is included, the need for a point of excess in every set for it to be consistent, how the threat of this point of excess leads to the necessity of a second count or metastructural state, and finally the means by which the relation between the count of belonging and the recount of inclusion, due to the point of excess, has distinct, political implications.[1]

Powerset axiom (Meditation Seven)

Meditation Seven commences by re-emphasizing the issue of belonging versus inclusion in relation to the traditional ontological pairs of one and many and multiple and part, distributed through the long debate as to the relation between unity and totality. Set theory's contribution to the dialectic of whole/part and one/multiple is that it suppresses both sides in favour of a sense of the multiple that 'consists from being without-one, or multiple of multiples' (BE 81). The result of this is that sets and multiples can be shown to be absolutely consistent immanently, without the use of such ideas as unity and totality, thus avoiding all the logical impossibilities of transcendental thought. Badiou realizes that set theory is a means by which immanence can be made consistent precisely due to its rejection of transcendence because being is-not.

Central to this innovation is the observation that there is no meaningful distinction in set theory between multiple and one, part and whole, rather we can speak solely of indifferent multiples of multiples.[2] Yet set theory does recognize another dialectical relation between multiples, that of belonging and inclusion. Belonging, defined here, is that a 'multiple is counted as element in the presentation of another multiple' (BE 81). Note the care taken in this definition. Belonging does not mean an element counted as part of a set, as this would tend to fall into the part/whole impasse that Badiou is studiously avoiding. Belonging is not a part that is counted in the whole but a multiple counted in another multiple. Inclusion, by contrast, is a multiple that is a sub-multiple of another multiple.

This discussion of the relation of multiples to multiples in terms of belonging and inclusion all depends on the powerset axiom we mentioned earlier. What the powerset or subset axiom says is that for every set there is a set of all its subsets or a subset of all its parts. If we take the void set as a collection that is empty {Ø}, we can say that the cardinality of this set, its size, is 1. The set is composed of the collection of no things. Nothing belongs to this set and this nothing is included as a named nothing {Ø}. If the cardinality of the void set is 1, however, the power of this set is 2. While only one element belongs to the set, the name of the emptiness of the set or {Ø}, two elements are included in the set, the name of the void, and the void itself which is always included in every set. The size of the smallest set is 1, but the power of the smallest set is 2, due to the difference between counting what belongs and what is included, or multiples as multiples and multiples as parts of or subsets of other multiples.[3]

What this reveals to Badiou is a point of excess typical of all sets due to the powerset axiom. The powerset axiom demonstrates that the number of elements included in a set is always greater than the number belonging to a set. If the cardinality of the void set is 1, its power set is 2 and so on following the order of 2^n where n is the number of elements in the set. This point of excess wherein a set includes more elements than it is composed of in a manner that rapidly expands into the realms of the uncountable, or the 'parts' always exceed the 'whole', is a simple concomitant of taking a set to be a collection not a fusion due to the axiom of separation. What set theory is effectively saying is the inside of every set is always larger than the outside in that there are more elements included in a set than belong to it. The only way this apparently impossible discovery can be possible is if we accept that sets are collections not fusions because then we do not presuppose that sets are wholes made up of parts and so we do not have to participate in the perfectly rational ban on a whole ever being smaller than the number of parts it contains.

We have to be clear and say that the powerset of all subsets is not a part of the set, it cannot be in any case as it is bigger than the original set, but an entirely separate way of counting the same set. Speaking of any set whatsoever as α Badiou says of this additional powerset count that this

> second count, despite being related to α, is absolutely distinct from α itself. It is therefore a metastructure, another count, which 'completes' the first in that it gathers together all the sub-compositions of internal multiples, all the inclusions. The power-set axiom posits that the second count, this metastructure, always exists if the first count, or presentative structure, exists ... [T]he *gap* between structure and metastructure, between element and subset, between belonging and inclusion, is a permanent question for thought ... What is sure ... is that no multiple α can coincide with the set of all its subsets. Belonging and inclusion ... are irreducibly disjunct (BE 83–4).[4]

From this we can gather the following conditions for what Badiou is going on to consider, which is the excess of the situation or how an event occurs by virtue of being. Firstly, the second count, that of the included elements, while seeming to be a stabilizing factor in that it is a total count as metastructure, is actually destabilizing because each time you take account of the elements of a set you find that the parts exceed the size of the whole in a manner that proliferates very rapidly to large, in fact uncountable, numbers. Second, while the presentative belonging is primitive, there is no presentation without representation or you cannot determine what belongs until you have counted what is included. This reverses or presents a chiasmatic loop as regards ontological precedence in that the elements which belong are not that from which the set is composed, but rather the set is composed of those elements which are included. Yet of course it is difficult to think of included elements unless first of all you have the basic belonging elements. How can you have subsets if there is no set? This conundrum reminds us that these elements are all the same, they are all multiples of multiples, so the issue is not one of ontological presence but of relationality due to the mode of the count. Finally, in all modes of ontology there is a gap, a difference, between essence and existence, Being and beings and so on. In Badiou's ontology there is also such a 'gap' but note it is of a profoundly different order. There is, as we have said, no difference ontologically between multiples that belong and those that are included. So, the 'difference' between a multiple that is included and one that belongs is an indifferent difference. Indifferent because there is no qualitative difference between the two multiples. The same multiple could be counted differently so that ontologically the multiple in question is the same, meaning the difference between a multiple as belonging and as included is indifferent, literally there is no difference in terms of the multiple as such. The final point here is that, due to the powerset axiom which states no multiple can be the same size as its subsets, the process of counting which seems to gather together a stable state always has included within it as a process an excess, namely that the size of the parts of any set is greater than that of the set itself, or there are always more multiples included in a set than belong there.

Point of excess

The first point taken from the powerset is that, counterintuitively, the parts of a set exceed the whole. A second issue must now be addressed which is that it is 'impossible to assign a "measure" to this superiority in size' (BE 84). This appears odd, considering that for each set we can give a number to its power. However, if we can accept this to be the case then 'the "passage" to the set of subsets is an operation in *absolute* excess of the situation itself' (BE 84). So, considering that we know that the power of the cardinal 1 is 2, that the power of the cardinal 2 is 4 and so on, why must it be that we cannot measure the superiority of the power relative to the cardinal as regards their size? For Badiou, the point of excess is innumerable for the same reason that we found in Aristotle that the void is innumerable: namely it has not been presented to the count. This makes simple sense in that if the subset is larger than the initial set and the initial set is due to belonging as pure presentation to the count, then the set of the subsets, which is not present to this count, cannot be counted within this particular set. Not being present to the count literally means you cannot be counted here, so in this sense the powerset, in that it is not presented in the count of the set it is the power of, while we can always assign a number to it, it is always beyond measure relative to the set it is the power of. So, if we say the power of 2 is 4, we are counting the power as part of the next set up, which is 3. The power of 2 is counted in 3 but the power of 3 is not counted in 3 but in 4 and so on. The size of the powerset is always bigger by definition than the actual cardinal size of the set. As the cardinality of the set is its measure, the powerset is immeasurable *within that set*. It is important to realize here that we are not actually talking about size per se, but the ability to be counted, which we tend to call size and measure. Thus, Badiou says, it suffices to show that the powerset of any set contains at least one element that does not belong to the set, in that it was not present for the count, to call it immeasurable.

As we saw in relation to the void, the size of the set was 1, one could speak of a collection composed of no elements, but the power of that set was 2. It included the void, and the singleton of the void. In that the void cannot be presented, included in every subset in fact is an element that cannot be counted but which is included in the re-count. Every stable state, every power-set metacount, not only includes at least one element which can never be present to the count, but in fact is ultimately founded on that element, as we shall see when we consider the axiom of foundation. Strictly speaking, as every powerset count includes the void and the subset of all multiples, neither of which can be counted in the set as such, every set has at its lower and upper limit, an immeasurable element. That every set has two immeasurable elements, the void and its upper limit of sub-multiples, we will discover, is the basis for Badiou's entire idea of structure as facilitator for event in *Logics of Worlds*.

We can accept the immeasurability of the powerset ontologically after Aristotle, but we can also prove it formally and in doing so unearth further impressive qualities of excess in particular as regards self-predication. Here is Badiou's argument. We start with a set α and we determine as β all those elements of a set that are not members of themselves. Here, yet again, we invoke Russell's paradox by defining those elements which belong to themselves as those 'which do not present themselves as multiples in the one-presentation that they are' (BE

84). These ordinary multiples, as Badiou calls them, first of all belong to α and second do not belong to themselves. Why make this stipulation? Well, if the element in question belonged to α *and* to itself, then it is impossible for α to be a counted element, as to be counted it needs to be re-counted and a set made up of a multiple that entirely belongs to itself is a set without any subsets. If a set belongs to itself, then it is not a multiple of a multiple. This means it is not included in any other set nor does it have any other multiples included within it, as these would be multiples of other multiples and in a set which belongs to itself all 'elements' must be of this set alone. There is a name for such an apparently impossible set which entirely belongs to itself, it is called the event, but we are some way off from being able to prove that. We can now call an ordinary subset, a subset of all elements that belong to α and do not belong to themselves, γ or powerset. What Badiou wants to show is that γ cannot belong to α even though this sounds entirely counterintuitive, as γ is made up entirely from the elements of α all of which belong to α by definition. If, he says, γ did belong to α then one of two things would be true logically. First, if we take γ to be ordinary then we say that it does not belong to itself because that is the definition of every ordinary set excluding the event. If we stipulate this and we insist that γ belongs to the ordinary subset of α as it must if it belongs to α, because the ordinary subset of α is simply all the elements that belong to it, then γ, in belonging to α, by default belongs to itself. Yet we have just said no ordinary set or subset can belong to itself.

At this juncture it is worth clarifying that, contra Russell's paradox, ZF set theory allows for a multiple to belong to itself, that is the multiple {Ø}, and indeed it needs such a multiple for the axiom of foundation due to the void.[5] So, Badiou suggests, we could accept that if {Ø} is possible as a self-belonging element, then generally it is always possible that such a thing as γ, a set in general that does indeed belong to itself, exists. If it did, it would not be ordinary, nor would it be {Ø} because we are speaking of any such set in general not a specific set named 1. Instead, it would be errant, impossibly in excess of the situation, an event. This sounds credible, after all γ is a very unusual set as we have seen in terms of its make-up, not least that its parts exceed the very whole from which they are selected. Yet if it is evental, in that γ belongs to γ, which is the ordinary subset of all the elements that belong to α, it has to be, by definition, ordinary. Clearly then if one insists that γ belongs to α you have to also insist that it is ordinary, in other words belongs to α and not to itself, *and* evental, it also belongs to itself. This is certainly impossible, and so the only other alternative is the conclusion that, amazing as it sounds, the ordinary subset of all the ordinary elements that belong to α does not and cannot belong to α itself as if it did it would no longer be γ, the ordinary subset of α, but something else we will come to call the event. Badiou concludes after this sequence of classical logical reasoning: '*no multiple is capable of forming-a-one out of everything it includes* . . . Inclusion is in irremediable excess of belonging' (BE 85).

We are now firmly lodged at the point of excess, proven by the simple contradiction that no multiple can be ordinary and evental at the same time, by which read in the same count. For remember no multiple is any way different from any other in essence. As all multiples are indifferent in terms of quality, the only way they can be counted is due to the axiom of separation by virtue of relationality. This means that every multiple can be ordinary and evental, depending on the nature of the count of their relationality as regards belonging and inclusion. In *Logics of Worlds* this will be formalized as the manner in which evental multiples are presented in a world to which they cannot belong due to their being multiples which are self-predicative and thus ontologically impossible. To reiterate, for set theory to retain realism it needs to reject Russell's paradox as regards the constructible nature of multiples only under the ban of ever presenting any multiple that is self-predicative. In other words, the seeds of the final construction of the idea of the event in a book to be written in two decades' time, are sown here by the various stipulations Badiou has to make as regards the nature of the multiple due to self-predication in order for his ontology to be both consistent and capable of retaining materialism or the void as real. An additional point is implied here although it will only be confirmed later. The process of re-presenting or re-counting which is a meta-structure – later Badiou politicizes this by calling it that which forms a state – is always founded on the necessity of the point of excess. In a moment of logic rather familiar to the philosophy of difference since Heidegger, every self-present state is founded on at least one element which is present as excessive to the self.[6]

Void as name

Badiou explains that it seems self-evident that if the void is not presented to the count, and thus can never belong, then it cannot be included either. If it is not counted how can it be re-counted, if it did not belong how can it be included, if it was not in the set how can it be a subset of that set? Yet we have already stated that the void is included and has to be for the powerset to function. Badiou now goes on to establish this more firmly on two counts. First, he says, the void is a subset of any set so is universally included. This is because all sets are composed in order from the smallest set upwards. That smallest set, which has the cardinality of 1, is, as we saw, the empty set which is a collection of no elements {Ø} or the first single unit. Set theory calls this the singleton. Second, the void set possesses a subset, which is the void itself which is included in every set to which it cannot belong. In that all sets are constructed up from the first, void set or {Ø} and that set includes the void as such as well as the name of the void, every subsequent set, which is in fact every set that exists, admits to the void in terms of its name, {Ø}, and in terms of the fact that it must always include the void as such. For Badiou this defines 'the omnipresence of the void. It reveals the errancy of the void in all presentation: the void, to which nothing belongs, is by this very fact included in everything' (BE 86). This point needs further clarification I suspect.

Badiou expands by saying 'if the void is the unpresentable point of being . . . then no multiple, by means of its existence, can prevent this inexistent from placing itself within it' (BE 86). If there is nothing to prevent the inclusion of the void or the inclusion of any inexistent element in any set, this does not quite mean that the void is always included, only that if it wants to it can be. Badiou explains further that the void cannot be counted as one, but it can always be subject to inclusion 'because subsets are the very place in which a multiple of nothing can err, just as the nothing itself errs within the all' (BE 86). From this we can take the following, that the nothing of being as inexistent to facilitate the count of what there is, can be said to be the void in terms of its being an inexistent multiple which facilitates the count of what is included in what there is. In the two instances it is the same nothing and the same set and the same multiple, but it produces different effects when applied to the count of belonging and the recount of inclusion. From this we can state that if there is nothing to exclude the void from the count, in that the count depends on the void's not belonging to the count, then if there is a count there needs to be the void and if there is a count there is also nothing to exclude the void, thus the void is omnipresent to every count. Badiou uses the logic of *ex falso sequitur quodlibet*, in keeping with his classical logic, to test this point concluding that the void set not only *can* be included in every multiple it is included in every multiple.

Badiou ends this rather complex, but necessary, set of proofs with the statement: 'The void is thus clearly in a position of universal inclusion' (BE 87), before adding that this also means, by definition, that the void, which has no elements, does still have a subset. The proof is disarmingly simple. If you accept that every set includes the void multiple, and as there is actually no difference between sets and multiples, if the void is a multiple it is also a set and thus said set must include the void as its multiple. Leaving us with the troubling conclusion 'Consequently, Ø is a subset of itself: Ø ⊂ Ø' (BE 87). (Note this is not the same as saying that it belongs to itself.) I hope I have prepared the reader in part for this enigma, in particular by my emphasis on indifference as content-neutrality, the nature of a set as a collection, the lack of ontological difference between set and multiple and the nature of the difference between belonging and inclusion. All of which points to the 'real' nature of sets, as opposed to the misrepresentative idea, rehearsed by Badiou here, that inclusion in a set somehow means being inside a set as is often shown by using Venn diagrams to capture the nature of sets. A practice widely regarded by mathematicians as profoundly misleading by the way. If we think of a subset as inside a set then we might look at the inclusion of a void as filling the void set with something, itself basically. But it is only belonging which 'fills' presentation. So, although it would seem absurd initially to speak of the void as a subset of itself, this is only if we think of void and sets in general in terms of the old logic of classes and properties.

Rather, if we say that the void is included in itself, is a subset of itself, this is the same as saying anything is a subset of itself which is, after all, the basic law of the set as collection: axiom of separation. Every set has at least one subset which is the set of all the elements which are included in a set. This is as true for the void set as it is for a set with three things in it or a million. At this juncture, speaking of the powerset or subset axiom, Badiou then explains that if the powerset axiom pertains to the void set, {Ø}, which it must for it to be a set, then we seem to admit a new contradiction. The powerset axiom presents the set of the subsets of a set, which means the set of everything that belongs. Yet if we speak of the void set, when we count what is included in a set, through the powerset axiom, we are forced to count only one element, the void, which is included in every set including the empty set, and then say that the void belongs to the void set. This seems to break the rule we have been promulgating here by saying that the void belongs to the powerset of its subsets. There is nothing else in the set so that nothing must be the thing that belongs, as there can be no inclusion without belonging. So, if the void belongs to the void set, in which way can this be said to be true? Badiou says: 'The set to which the void alone belongs cannot be the void itself, because *nothing* belongs to the void, not even the void itself . . . It would be the *name* of the void, the existent mark of the unpresentable. The void would no longer be void if its name belonged to it . . . the void cannot make a one out if its name without differentiating itself from itself and thus becoming a non-void . . . Consequently, the set of subsets of the void is the non-empty set whose unique element is the name of the void' (BE 88).

The separation between being as such, shall we call it, and its name is now complete in terms of the difference between the category of belonging and being included. You can include the void, through naming it, only as that which is present as non-belonging. You can prove this using the logic of metaphysics, as we have done, or that of set theory, as we have done, or also that of standard logic, which we have referred to here and explained in part. In other words, the transmissibility of void being exists across three differing, rigorous forms of discursive communicability and all because of the indifferent nature of being as void. As we can see, if the name of the void belonged to it, it would no longer be void as the void is in-different. To make a one out of the void it must be differentiable and this is not possible for the void, which has no quality and no presence so simply cannot enter into the logic of differentiating separation. Yet you can differentiate the void by including its name. When you name the void you can count it, when you count it you negate it, because you differentiate it and a differentiated being is a non-being, because being is defined as in-different. That the void is also indifferent is another matter. Whether you take the void as radical in-difference, impossible to present to the count of differentiation, or as indifferent due to content neutrality when it is presented as a multiple in the form of its name, either way the point is clearly made: the void as non-belonging and as included in the count depends in its entirety on the axiomatic method of set theory which insists on the indifferent fundamental nature of every multiple.[7]

Four kinds of one-ness: one, count-as-one, unicity, forming-into-one[8]

The seventh meditation comes to a close with a consideration of four meanings for the term 'one'. The four aspects of the one – one, count-as-one, unicity and forming-into-one – are all designed to negate the idea of the one as something transcendent, foundational or preceding the count. Having, therefore, been schooled by set theory to think differently as to what constitutes the many, a collection as multiple of multiples, we are now moving towards thinking differently as regards the one of said multiples or what is traditionally taken to be a set.

The one, as such, is-not. This axiom disallows any future errancy into suggesting otherwise. The other three senses of the one are based on this first sense and so in no way can they allow the one, which is-not, to transform back into the one as unified, transcendent being. As we now know, the one is nothing more than an effect of the count, while what is actually counted is always multiples of multiples, which naturally can never be the One. What is counted as one in set theory then is never the One as such, the One is outside any count, but rather a multiple of multiples or a set. This being so, Badiou explains, 'It is thus necessary to distinguish the *count-as-one, or structure*, which produces the one as the nominal seal of the multiple, and *the one as effect*, whose fictive being is maintained solely by the structural retroaction in which it is considered' (BE 90). Retroaction is a rather crucial part of the whole structure. The one can only be counted by being assigned a fictive name subsequent to the presentation to the count of a multiple, so that it can be counted-as-one in such a way that it founds presentation by not being presented to presentation. This is because, as we have seen, the one as void is the pure presentation of presentation as such. There is, then, no one as such, but only one produced subsequently due to the count. Badiou calls this the 'fictive one effect' and concedes it is a dangerous designation, especially as regards the void set, because one might mistake the act of naming for the thing itself. When I say that Ø is the void, one might erroneously deduce that I am simply classifying the one-object by giving it a name. In fact, I am allowing the one to be presented by naming it as void, as present as the non-presentable in every presentation. In this way I circumvent both the problems of predicative objects and the pitfalls of the solutions proposed to these problems by modern philosophies of various kinds.

The count-as-one allows for the possibility of unicity, a possibility denied the one as such by Badiou. Unicity seems, on the surface at least, to begin to contravene our insistence on indifference at every stage of Badiou's ontology thus far. Surely uniqueness demands the assignation of qualities unique to or in unique combination within a very specific thing? Yet this is not the case as we saw. Unicity is indifferently unique precisely because, as Badiou says, unicity is not a being. The assumption of a unique being which is then assigned a predicate, this is a unique thing, is incorrect in this context. The void cannot be unique because unicity depends on differentiation and the void is in-different. This is an essential point as it allows us to contravene one of Derrida's main contentions as regards the possible existence of an event. For him, something truly unique would be impossible to recognize and so difference is never absolutely different.[9] What Badiou brilliantly proposes here as he builds the groundwork for his theory of the event, is that unicity is a quality of all multiples and not a pre-requisite for what we shall call an absolute, nonrelational difference or event. Badiou insists that unicity does not belong to being, being is-not so cannot be unique, but rather to the realm of same and other, which are effects of structure.

'A multiple is unique inasmuch as it is other than any other' (BE 90). Unicity appears in this instance as the result of indifferent, Saussurian comparative difference. So, too, the unicity of the name of the void. When we name the void we do not name its unicity as being, but simply its unicity within the count: 'the name of the void being unique, once it is retroactively generated as a-name for the multiple-of-nothing, does not signify in any manner that the "void is one"...' Rather, all it shows is that given the necessity of this fictive nomination, 'the existence of "several" names would be incompatible with the extensional regime of the same and the other, and would in fact constrain us to presuppose the being of the one, even if it be in the mode of one-voids, or pure atoms' (BE 90).

Forming-into-one, Badiou freely admits, is really a mode of count-as-one and determines that any counted one can immediately form the basis of another count and that of another count ad infinitum. This is because the count is 'not transcendent to presentation' (BE 91) so that as soon as the one is presented it is then a multiple of another count, oscillating endlessly between a one (set) and a multiple in an infinite amount of cumulative counts. In terms of representation if the one is named Ø then the forming-into-one, the use of the one as a multiple included in another count rather than an element of the initial count, should be represented as {Ø}. Here we encounter a fundamental interaction between the powerset axiom and the axiom of replacement. As we saw the powerset axiom states that there is always a set of all the elements included in a set, which is larger than the number of elements belonging to a set because the powerset always counts the void as one, whereas the void can never belong and so submit to the initial count. This means that in the void set the contents of the set are {Ø}. No element belongs to the set, but that non-belonging element is still included. Another name for the void set in set theory in relation to natural numbers is 1 or, as Badiou calls it here, the singleton: 'The set {Ø} is thus simply the first singleton' (BE 91). This law of the singleton, which is the structural minimum of set theory that disallows infinite regress – there is no set smaller than the singleton and the singleton does exist – is supported by the axiom of replacement or of the total indifference of Ø. This axiom says that any element can be replaced one-to-one by any other element and the set remain the same.

Thus, instead of talking of the void we could talk of δ, abstract notation for any element whatsoever, and say that, due to the powerset axiom, if δ exists then {δ} also exists. In other words, we can use the peculiarity of the void set and its naming to derive the singleton existence of any indifferent thing. This is just another way of stating that because the one is-not, the multiple, here in the singular, is. To clarify the exceptional nature of the reasoning combining the axiom of the power set and replacement Badiou is able to say that if Ø inexists then {Ø} exists and if {Ø} exists then {δ} also exists. Yet if you look at what is precisely being said here it is rather particular. If Ø does not belong as it inexists this means it is not presented. Yet even though it is not presented it can still be included in the power set count. This results in the first singleton: nothing can be smaller than the set {Ø}. As this singleton exists then, due to the axiom of indifferent replacement, you can replace Ø, which does not exist and so does not belong and is not presented, with δ which does exist, does belong, is presented and so can be counted. And if δ exists then due to the powerset axiom so does {δ}, meaning you can derive the existence of the singleton multiple from the inexistence of the singleton void.

We can now close by noting that the process of forming-into-one not only founds the minimum level of the first singleton, but it is also the basis of infinite singletons each unique, each founded by forming into one the name of the singleton based on the previous singleton. So {Ø} is formed into one out of Ø, then {{Ø}} is formed into one out of {Ø} which is not the same, and so on. We will cite Badiou in full here:

> To conclude: let's note that because forming-into-one is a law applicable to any existing multiple, and the singleton {Ø} exists, the latter's forming-into-one, which is to say the forming-into-one of the forming-into-one of Ø, also exists: {Ø}→{{Ø}}. This singleton of the singleton of the void has, like every singleton, one sole element. However, this element is not Ø, but {Ø}, and these two sets, according to the axiom of extension, are different. Indeed Ø is an element of {Ø} rather than being an element of Ø. Finally, it appears that {Ø} and {{Ø}} are also different themselves (BE 92).

The significance of this observation for Badiou is that 'the unlimited production of new multiples commences . . . starting from one simple proper name alone – that, subtractive, of being – the complex proper names differentiate themselves, thanks to which one is marked out: that on the basis of which the presentation of an infinity of multiples structures itself' (BE 92). What we are proposing is that the forming-into-one plus powerset axiom allows one to construct an actual infinity of multiples, due to one-to-one correspondence thanks to the axiom of replacement and the addition of the axiom of extension, which says that in each case the new multiple is different. Remarkable though this is it is even more astonishing to note yet again the debt Badiou pays to indifference. Each of these sets is in terms of 'content' apparently the same {Ø} {{Ø}} {{{Ø}}}, yet according to set theory and its dependence on collection rather than fusion, each is formed-into-one from the previous one, so while each is made up of the same core element, the relation of belonging is different. This procedure of the forming-into-one reinscribes the primitive law of logical indifference which is also the basic law of logic and set theory: a multiple is determined by relation not by content.[10]

To sum up, we can see how every multiple is composed from the initial non-belonging of the void set as singleton and the forming into one of a set derived from this first, minimum set, a set with no elements belonging to it meaning it cannot be further subdivided. Using the axiom of replacement along with the law of forming-into-one an infinite number of actual singleton sets can be derived due to a one-to-one correlation or axiom of replacement plus that of extension. Each set has a different 'name' yet each set is composed of the same, single, 'element', the void. Here then we have the basis of Badiou's universe: the founding of multiples on the void, the structuring of multiples of multiples emanating from that first singleton set, extending out to the point of infinity. This is set theory sutured to philosophy. The problem of the one and the many is solved by the count-as-one due to the one which is-not via the multiple of multiples. The count-as-one solves the problem of infinite regress, no set is smaller than the singleton, and of bad infinity, the proliferation of infinite singletons is actual, it can be 'counted' using basic one-to-one correspondence. The specific proofs of these two positions are to come when we look at the axioms of foundation and actual infinity. It is perhaps needless at this stage to point out that it also marks the dependency of set theory on indifference. If I may be so bold this is Watkin's universe: the in-difference of being, the content neutrality of multiples and the indifferent difference of each singleton to that which precedes it each of which has a name, a unicity, that is non-unique because it is a contentless name of a content-neutral multiple.

The state (Meditation Eight)

In Meditations Eight and Nine, Badiou makes his most controversial assertion, namely that just as there is a metaphorical relation between the terms of set theory and politics so too there is a stronger relation which allows one to present a political philosophy justified mathematically in some way. To shift the terms of the debate in this direction he starts to call structured situations 'states' with all that implies politically. It would appear that most readers of Badiou are perfectly willing to accept this relation. Hallward takes it as read, and even though Norris realizes that to the intended audience of his study, philosophers from an analytical background, such a leap from formalization to politics would be highly suspect if not risible, it is apparent from Norris' overall tone that he accepts Badiou's politics and is happy enough to see mathematical formalization used to promote beliefs that he appears to share with Badiou. Indeed, more widely, it is Badiou's politics that have had the most lasting impact so far within the philosophical community.[11] Yet for our own part we have set ourselves a more ascetic program, in keeping with Badiou's own fearsome intellectual monasticism. We have decided to consider *Being and Event* in terms of what is justifiable formally due to set theory, philosophically in terms of the larger problems of metaphysics since the Greeks, and finally as regards the manner in which *Being and Event* presents Badiou's complex sense of indifference as a positive avenue for future thought. On our terms, therefore, we need to ask the question is the state, the term under debate here, a justifiable one as regards set theory, metaphysics or indifference? In particular, is it not the case that in deciding to call structures states, Badiou makes structures appear more political than they actually are? For example, when he returns to structure in *Logics of Worlds* he replaces, for the most part, the term 'state' with 'world' which immediately sounds less politicized even if the book as a whole is openly political in intent. What we need to show then is that the political aspect of Badiou's work goes further than analogy between formal states and real-world states, that it is not dependent on the use of the term 'state' alone and that we have not made Badiou's politics communicable simply because they accord with our community's left-leaning tendencies without any clear rational and philosophical base. We will not be able to answer to these issues by reading *Being and Event* and the works from the early to mid-period of Badiou's works alone, *Logics of Worlds* has to also feature prominently, meaning we will not answer the question as to the precise nature of Badiou's politics in this first volume.

Accepting that let us at this stage consider the proposition which is at the root of these two related meditations. Badiou says that due to certain threats inherent in the count of presentation, belonging, it is necessary to have a second count which determines elements which are included. He calls this representation. There are, in this way, always two counts. If the first count determines the structure of the situation, the second therefore is a meta-structure. If we call the first count the situation, remember multiples count as one only specific to a situation, the second is the state of the situation. The two counts are clearly relational, albeit in specific ways we will explain. Finally, because of this terminology of the state of a situation, it is legitimate to look at other theories of the state, in particular here Marxism, and construct a more robust political philosophy than has been possible up to this point, by virtue of the rigorous nature of set theory. This is the thesis. Our question is, why must there be two counts, and why should this double count be by definition political?[12]

So far in my studies of set theory I have found no significant consideration of the count of the count, the state of the situation, the meta-structural and so on. Indeed, in terms of how I have presented set theory, the re-count, in as much as it is not indifferent but is defined by classificatory qualities, is not directly relevant to set theory as such. Norris is also unable to axiomatically prove the recount theorem. Rather, he proceeds operationally by saying naturally set theorists double-check their findings against the acceptable norms, using Kuhn's definition here of mathematical practice, and so in this way they re-count (BBE 87). I find this unconvincing and indeed perhaps problematic as it undermines the axiomatic retroactive method to some degree. So, apart from the bare fact that mathematicians double-check their results against accepted norms, I can present no direct relation between set theory and a second count. More than this, although Norris says that naturally because terms like state are politically redolent it is natural one should use them politically, we will call this the analogy argument and suggest it is also suspect. I have not come across any serious tendency in set theory to use the term state. And as state is not a central term of metaphysics, and more than this it seems to take us far from the indifference of presentative situations, there appear to be strong reasons for rejecting the state of the recount as a significant part of Badiou's work.

We will not be so hasty, but I will say that there is no rigorous evidential support for the relation of representation to political states and so this side of the debate I will leave to others taking their lead from Gillespie. In answer to his own question 'Is there a state for set theory?' he concludes: 'Set theory, and consequently being itself, is a fundamentally incomplete situation. It lacks a state that can count it as one' (Gillespie 64–5). What this means is that set theory as ontology is not sufficient for a consideration of the state. The wider implication of this valid point is first, that the state as a concept cannot be completed until *Logics of Worlds* presents a new mathematical condition to supplement set theory (there is a state in category theory albeit one Badiou now calls a world) and second, that the theory of the event, which is due to the excess of the state in relation to itself, is not provable from within ontology.[13] The second point is accepted by most critics but I am not sure the first necessarily is. As far as our study is concerned, there is nothing intrinsically political in Badiou's ontology, he would agree with this, and the suturing of philosophy to politics at this juncture is, on our terms, unconvincing and premature. Instead, what we believe is that Badiou moved from the idea of set theory as a re-counted state, over nearly two decades, to the application of a new mathematical system, category theory, to describe meta-structural states as worlds. As he did so he justified the second count, not in terms of sets but categories, and recalibrated the rational justification for his politics, which do not change in themselves. Any reading of Badiou's politics that does not concede this point is effectively twenty years out of date.

Threat of the void

Badiou deduces the necessity of a re-count, or better re-presentation, from the axioms of set theory even if, strictly speaking, there is no concept of the re-count in set theory. This occurs due to two phenomena, the threat of the void and the point of excess. As we already saw every set includes more multiples than actually belong to it. For a set with several multiples it is obvious that the number of submultiples, Badiou tends to prefer this to subsets, will have a cardinality larger than the initial set. Cardinality, remember, is the numerical size of any set. There are simply more parts to a set than there are elements because parts can be combined in different combinations to make different subsets which are new parts combined from the same elements. Yet, we also saw that even the empty set has a singleton element which is {Ø}. In that the void set is included in every presentation, and cannot be present to the count, there is always at least one multiple 'more' in every set included but not counted, the void. Badiou explains, quite rationally, that for any multiple-presentation to occur, the void element cannot be fixed in the presentation. By fixed here we mean presented as counted following the sequential numerical ordering of how you determine the cardinal number of every set in terms of well-ordered ranking of its ordinals. If the void were counted, then presentation would present the unpresentable and would immediately collapse. This is the first danger of the void: that you may try to count it. This does not threaten the void itself which is not negated through being counted, only the consistency of presented multiples.

There is, however, a second danger of the void which pertains to the presentation of presentation as such. As Badiou says: '*something*, within presentation, escapes the count: that something is nothing other than the count itself. The "there is Oneness" is a pure operational result, which transparently reveals the very operation from which the result results. It is thus possible that, subtracted from the count, and by consequence a-structured, the structure itself be the point where the void is given' (BE 93). If we put these two threats together we see that the void presents a local threat, you may try to count it, and a global threat, what is actually void is pure presentation as such and structuration transparently reveals this. So, if you come at a presentation in terms of 'parts' of 'wholes', the void will make its presence known. And in terms of structure, in that structure is nothing other than relation, and the void of indifferent presentation is non-relational, again like the attempt to fix the void, structuring the void is the very negation of structure as such. The void then is the basis of being able to count as one and to count ones as one, yet it also threatens the consistency of these two senses of the one.

It is the second point that detains us as we are now primarily interested in sets of ones taken as consistent ones in themselves. As Badiou says, "*it is necessary that structure be structured*" (BE 93). However, due to Gödel's incompleteness theorems, it is now widely accepted and communicable that no structure can be validated by taking the axioms of a system and applying them to the whole structure. Another version of the self-predication paradox effectively. This is what is meant here by meta-structure: the philosophical justification of entire mathematical systems using the tools of those systems alone.[14] I have already raised my own objections as to the meta-structural justification of the state as re-presentation and Badiou admits that his thesis appears to be *a priori*. Yet, he says, there is an intuitive, by which read empirical, justification for it that 'everybody observes and which is philosophically astonishing' (BE 94). This justification is that while the 'being of presentation is inconsistent multiplicity ... despite this, it is never chaotic' (BE 94). There is a hastiness here in that until the publication of *Being and Event* it was not communicable and transmissible amongst us that the being of presentation was inconsistent multiplicity, just as it is hard to say that everyone would agree that what is presented is not a form of chaos. All the same the point is well made. Having accepted inconsistent multiplicity, which we have, it is surprising that we do not end up with the double impasse of infinite regress plus bad infinity as that is where everyone else fetched up. We have seen the axiomatic reasons as to why we do not succumb to infinite regress already, axioms of foundation and the singleton of the void {Ø}, but we have not yet fully considered the other alternative, and so we can now begin to feel the shift in emphasis from being towards event due to the imminent axiom of infinity. The second count, if we call it that for the time being, is really another name for the axioms that stop the basis of multiplicity overwhelming multiplicity immediately so that it is fundamentally useless to us. Badiou, trying almost too hard to convince blurts out: 'All I am saying is this: it is on the basis of Chaos not being the form of the donation of being that one is obliged to think that there is a reduplication of the count-as-one. The prohibition of any presentation of the void can only be immediate and constant if this vanishing point of consistent multiplicity ... is, in turn, stopped up, or closed, by a count-as-one of the operation itself, a count of the count, a metastructure' (BE 94).

Let me make a few basic points here in terms of Badiou's insistence on the second count. First, the void cannot be accidently counted as present. Second, the transparency of structuration, the presentation of pure presentation as such, cannot be the global effect of the count because it is not present and you cannot present the non-presentative basis of presentation (by present here we mean count). Third, if you accept the basic axiom being is-not then you can simply observe that what is present is not a pure chaos and accept there must be a second process which moves from being's inexistent inconsistency to existent and consistent multiples of multiples. Fourth, as we have seen already, the whole point of Badiou's ontology is to solve the problem of the part and whole without recourse to infinite regress and bad infinity. So, fifth, there must be a stopping point for every multiplicity at the lowest level, to protect it from the pure void, and at the highest level to allow us to give the one thing I have avoided speaking of thus far: set or multiple specificity. The metastructure moves from the count of indifferent multiplicities to a specific count of these multiplicities in this situation. At which point, on the whole, said multiplicities cease to be indifferent and instead become differentially communicable. Which leads us to our sixth point, not expressed by Badiou but deducible from all he does say. While being is indifferent, its worlds are not. If the first count of presentation confirms again and again the reliance of Badiou on indifference, the second shows that from the perspective of the state, there is no indifference. Politically this is a valuable lesson. Finally, a seventh point must be made here. The meta-structural concerns that Badiou speaks of that negate the vanishing point of multiplicity etc. are nothing other than the axioms of set theory. This is important first to establish a clear demarcation between Badiou's ontology, being is-not, and the mathematical proof of this ontology. They are not the same. Set theory applies its axioms to specific situations to fix presentations precisely so they are consistent, count-as-one, yet based on the inexistence of the one. Put simply, set theory is metastructural, a state, a representation and, using a term he starts to apply to all states, a fiction.

In the passages that follow Badiou speaks of the haunting of the void, its anxiety, its danger, while he determines the 'veracity' of the 'one-effect' as 'literally the fictionalizing of the count via the imaginary being conferred upon it by it undergoing, in turn, the operation of the count' (BE 95). For much of Badiou's career until recently he has favoured, like a good Marxist, a withering of the state, a destructive debilitation of its effects. Yet what marks Badiou's politics out from traditional Marxism is his inability to favour a single revolutionary event and his acceptance that every revolution is followed by a re-count, a state, a betrayal. One could argue this using Derrida's logic that all purity is immediately impure once it is named, which is true of the event, but here the more powerful argument is that the logic of inconsistency, set theory, is consistent and transmissible. This would be true if it were not the case that set theory is not philosophy for Badiou. Set theory certainly is a state; it is the second count that performs all the duties I have just listed. Yet set theory is not a meta-structure per se. It is conducted under the auspices of Gödel's incompleteness theorems and interests Badiou as much in terms of its exemplary evental axiomatic process, still ongoing, as anything else. All of this pertains to Badiou's extensional realism, an area Norris gives an excellent reading of. As he says, ontology is an incomplete science because multiples are always multiples to a situation, immanent. Thus, for Badiou, there is a real which set theory operates on and if there is a real, then naturally, contingency rules. This disallows a final event of all events, orthodox Marxism. And it disallows all intensionalist language-game theories, because set theory pertains to real situations which exceed its transmissibility so that from time to time it has to present the communicably unpresentable and invent new axioms, something language-game theories never do.

Norris uses Badiou's realist extensionalism to mount a defence for his politicization of representation as a state. Let us trace his argument because it pertains directly to the role of indifference in Badiou's realism, his extensionalist set theory and, Norris says, his politics. For extensionalists, set theory axioms have their truth-value 'fixed by the way things stand in mathematical reality' (BBE 91). While for intentionalists 'truth becomes a matter of epistemic warrant' and there is no truth-maker, no reality, beyond that which guarantees truth. Norris then argues that Badiou's extensionalism is a 'strictly non-differentiating ontology of sets, subsets, elements, parts, members and so forth, which leaves no room for any kind of qualitative distinction and which therefore conceives their various orders of relationship in purely numerical terms' (BBE 91). How is this the case? Norris explains this when he describes the universalizing effect set theory has had on Badiou's Marxism. Rather than concerned with how things actually stand in a particular situation, set theory demands an abstract and universal set of laws that apply to all situations. So, that while multiplicity is immanent, multiple to a situation, it is indifferent as regards which situation. In effect, all situations are the same, a form of radical structuralism with the event added in to facilitate theories of sudden change. Norris now explains why this structural-mathematical approach facilitates indifference when he says of extensionalism:

> the operative sense of a term is fixed entirely by the range of those objects ... to which it applies or extends and not by anything specific *about* those objects (their distinctive properties, qualities or attributes) that marks them out as rightfully falling under the term in question ... an extensionalist conception of sets and their membership conditions is one that rigorously excludes any thought of whatever might otherwise be taken to distinguish potential or candidate members, and thus to determine which items qualify for inclusion (BBE 90–1).

This being the case, everything that is a multiple is determined by the laws of set theory, not only those things that clearly fall under its remit, and this means political states as well as numerical ones. If an element can be counted, then it is a multiple and set theory pertains. As Norris then convincingly argues, it is Badiou's realist extensionalism that allows him not only to apply set theory to politics, set theory is applicable to any counting situation in fact, but also to specifically define an idea of social justice based on the fundamental division in set theory between belonging and inclusion. Norris concludes that political interests in social justice 'can be best served ... only by a radically egalitarian and universalist outlook opposed to any form of identity politics, or any notion of difference ... ' (BBE 91). Badiou's politics, on this reading, are the politics of indifference based on the fundamental distinction between belonging and inclusion. We must now then investigate this division in more detail.

Belonging, inclusion and parts

The differentiating relation between belonging and inclusion, presentation and representation, is vital and marks indeed for many the most problematic element of Badiou's work which is how his ontology applies to the world. In as much as *Logics of Worlds* is designed to address this problem we can now say that the controversy has now been assuaged, still here the seeds of the issue are sown and we need to spend some effort cultivating them. In particular, Badiou is still justifying the idea of representation which he decides to call the state of a situation: 'that by means of which the structure of a situation ... is counted as one, which is to say the one of the one-effect itself' (BE 95). Having presented metastructure's necessity he now endeavours to define its nature. A metastructure or representation cannot be a bare count because then it would be indistinguishable from presentation, whereas difference from presentation is the whole point here. At the same time Badiou is not happy defining representation as the count of the count, not least because a structure is not a term of a situation so it can't be counted, we saw this in relation to pure presentation as such. 'A structure exhausts itself in its effect, which is that there is oneness' (BE 95) or, like being, you will know structure only by its works, what it allows to operate, without remainder.[15] Structure is not a multiple and it cannot be counted. This is a very important point: structure is relation and relation cannot be counted as it is the precondition of the count. It may be worth noting that if we are saying structure is a world in sheep's clothing then this position alters in the second volume where it becomes apparent that the meta-structural count of any consistent world is also an object of that world. That is twenty years away. Back in 1988 we are struggling. Meta-structure exists, Badiou asserts, but of what does it consist? He decides to tackle this from the other direction in terms of the void.

If you remember, meta-structure is a prophylactic against the threat of the void. Seeming to contradict his earlier comments, Badiou suggests that the void cannot threaten at the level of the term, as it is what is subtracted from the count, nor can it threaten the whole as it is 'the nothing of this whole'. If there is a risk of the void negated by representation it does not concern terms or wholes, neither of which can in fact admit of the void as they are based on the no-thing of the void. So, what does the re-count count? The answer is parts. Parts are significant here in terms of set theory, they determine the relation between belonging and inclusion; metaphysics, the whole-part impasse is the basis of Badiou's ontology; and indifference, in that a part is a suspended situation between terms and wholes that leads to indifference becoming a difference. Speaking of the count, there are as we have consistently said, no parts as all multiplicities are indifferent, consistent one-multiples. A part generates 'compositions out of the very multiplicities that the structure composes under the sign of the one. A part is a sub-multiple' (BE 96). We are still mired in difficulty, however. For example, if a multiple is presented as a composed multiple, so are all multiples, meaning it is a term not a part. If however this part then does not form a one, is not a term, then it is-not. So, it would seem at first that a part does not exist because either it is a term, or it is-not.

At this point, Badiou is able to return to orthodox set theory primitives. If a part counts as one, it belongs and is a term, if it is a composition of multiplicities it is a part and does not appear in the count because it is included and does not belong. Remember that there are always more elements included in a situation than belong to it so there are always elements that are not presented but can be represented. What precisely are we talking about here? For example, in the count 1-4, four elements belong. But many more than 4 elements are included for example the additional units 1,2 and 1,3 and 3,4, and so on, as well as, as ever, 0. These sub-sets, Badiou calls them sub-multiples, must be composed of presented elements, so they are naturally a second count, and they do not belong to the initial count because until you know what elements there are you cannot count their subsets. Hallward gives a more political reading. If the initial count is taken as citizens, then the re-count is grouping citizens into smaller groups, say for demographic control of voting procedures. We will come back to this in a moment as this is the only speck of indifference found within a state. Ironically, states are regularly accused of indifference yet Badiou's work shows that states cannot be indifferent: states are structures, structures are relations and relations are differences. Perhaps the easiest way to conceive of this is that what belongs is what there is, and what is included is how these elements are classified. It is at this point, and only at this point, that 'sets' are composed of elements that share in common *qualities*. Meaning for the record that until the eighth meditation, Badiou's ontology is entirely quality neutral, extensional and indifferent.

Badiou describes the universe we have lived in since Aristotle, that of difference, through the means by which representation, or the state, systematically negates the indifferent and quality-neutral basis of being. Up to this point we have gleefully detailed the necessity of being as regards indifference. However, the sad truth is that the final result, the final state of any situation, the worlds we all live in, is achieved through the negation of indifference represented here by the threat of the void as inexistent part and self-presenting whole. Badiou valiantly presents all the nooks and crannies through which the void could penetrate presentation and representation as well as arguing the possibility that there is no clear distinction between belonging and inclusion and in effect there is only inclusion. He toys with the problem of parts not belonging and so opening up the possibility of a void structure, in that the void is that which does not belong and of course the void set is always included in any set but never belongs. All of these sorties against the keep of belonging by the besieging problems of the inconsistent are lined up so that, by virtue of the axiom of excess, Badiou can first say yes, there is such a thing as inclusion because there are always more parts to a set than belong to it. And second more than this due to the multiple threats of indifferent void, not least the fact that Badiou's extensionalist set theory method depends on this indifference as Norris shows, the second count is essential: '"sub-multiple" – must be recognized as the place in which the void may receive the latent form of being; because there are always parts which in-exist in a situation, and which are thus subtracted from the one. An inexistent part is the possible support of the following ... the one, somewhere, is not, inconsistency is the law of being, the essence of structure is the void' (BE 97). This outcome would spell disaster and so metastructure is needed because 'metastructure guarantees that the one holds for inclusion, just as the initial structure holds for belonging' (BE 97). What this allows is that for any situation there is always a mechanism that counts as one this composition, blocking out the threat of the errancy of the void. The void threatened at the level of the term and of the whole, now it threatens at the level of the part because something included can non-belong and that which does not belong is-not. Thus the metastructure closes the gate with this moment of brilliant, circular reasoning: 'What is *included* in a situation *belongs* to its state' (BE 97). Every part is made into a one by the state, there can be no inclusion of the void, and then, at the end, all the parts are counted as one total part, echoing the initial presentative count.

> If the state structures the entire multiple of parts, then this totality also belongs to it . . . The state of a situation is the riposte to the void obtained by the counting-as-one of its parts . . . for both poles of the danger of the void, the in-existent or inconsistent multiple and the transparent operationality of the one, the state of the situation proposes a clause of closure and security . . . (BE 98).

As regards the in-difference of the void and the presentative indifference of presentation itself, the state imposes a one-effect which is the world in which we live. As long as we live in the world, according to this dictum, we live according to differences countable as ones. In the second volume we will revisit this issue and discover, surprisingly, that even in the worlds of pure relationality which replace what Badiou is calling here metastructure of the state, indifference has a prominent role to play, but for now we will accept that the primary effect of the re-count of every state is to negate indifference due to absolute relational difference.

Typologies of being

It is at this juncture that Badiou decides to introduce the idea of typologies of being that will prove to be rather important having an impact on central moments in the development of the event in *Being and Event* and featuring prominently in *Logics of Worlds* as well. Effectively what Badiou realizes is if there is an interaction or articulation between belonging and inclusion, then there are three types of being possible. Normal being is a multiple that is presented and represented. An excrescent being is a term represented but not presented. While a singular being is the opposite, presented but not represented. The event is impossible to conceive of without this typology of normal, excrescent and singular beings. Yet this also means that the event is deductible from set theory if you accept in set theory that there is a count and a recount, for the reasons just given, as this allows for different combinations of belonging and inclusion such that singular beings, in particular, exist, as singular beings take us to the possibility of there being events.

A normal term is by far the largest majority of all terms. Singular terms, as the name suggests, are rarer. Like a normal term they are 'subject to the one-effect' but they can't be thought of as parts as they are made up of elements that are 'not accepted by the count'. This means, significantly, that a singular term is undecomposable, in that what it is composed of cannot then be represented in a situation in a '*separate manner*'.[16] A singular term can never be called a part and as the state only counts parts because its count is always a recount 'an undecomposable term will not be re-secured by the state' (BE 99). Although a singular term exists, the state cannot verify its existence. The political as well as the ontological implications of this are of great interest to Badiou.

An excrescence is, as its name suggest, murkier. Badiou says 'it is a one of the state that is not a one of the native structure, an existent of the state which in-exists in the situation of which the state is the state' (BE 100). Ontology per se cannot possess excrescences, just as states cannot possess singularities. In that belonging is about presentation and nothing else, excrescences are state-constructions because it is not possible ontologically that anything be represented until it is presented. Again, there are political implications here but this time of a negative kind. As Pluth says: 'In a Marxist analysis of a bourgeois state, the state itself is said to be an excrescence because the state is a thing of pure representation with no real grounding in the multiples that are present in the situation'.[17] A second, more profoundly philosophical, point with political implications is made here by Badiou, although the means by which he secures his point are troubling. The reasoning appears to be that there is a clear difference between belonging and inclusion. Yet this difference is ontologically indifferent as there are not types of multiples, so the difference resides only in how you count multiples. Set theory insists on the existence of subsets whose excess allows for Cantor's proof, for example, of actual infinity. There are also innumerable other communicable benefits for mathematicians in axiomatizing the separation of sets from subsets. Yet ontology itself is undecomposable, which is another way of saying all multiples as such are entirely self-belonging. There is no gap in a multiple. This being the case Badiou says he is justified in suggesting the existence of a meta-structural entity, what he calls here a state, vouchsafed by ontology but by definition separatist, differentiated, and so 'different' from ontological belonging. It is provable that inclusion is just another form of belonging, but in terms of set theory, they are not the same.

The difference between belonging and inclusion not only allows a whole host of new possibilities to be communicable for mathematics, powersets, actual infinities and so on, but it also presents possibilities for philosophy. The first, as we saw, was the politicization of meta-structure in terms of the state. The second is the inference of a triad of types of being, normal, singular and excrescent. Now Badiou runs the tape backwards. If, he says, there are different types of being, then one of these types presents a problem for ontology, namely excrescence. It is perfectly feasible that singular beings exist. In particular, in *Logics of Worlds* we shall see that existence is merely the relationality of a being in a particular world. If such-and-such a being has no relationality in such-and-such a world, it simply means it does not exist in said world. This has no effect on being, just as being included in a world and thus existing has no effect either. It is, however, very different to say a term is included but does not belong. Ontologically speaking this is not possible.

Reflecting on excrescences, Badiou says what while ontology needs a theory of the state, it must remain state-less. If there were a state of ontology, or later a world of ontology, then the foundational difference between belonging and inclusion would break down. It would become necessary to speak of a rupture or separation within the pure multiple itself, resulting in the need for two axiomatic systems, that of belonging $\in$ and that of inclusion $\subset$. At this moment, ontology would collapse as set theory is based on 'the axiomatic presentation of the multiple of multiples as the *unique* general form of presentation' (BE 100). Badiou now says that because subsets must exist, this means excrescences exist, but because excrescences cannot exist ontologically, without ontology ceasing to be what it is, then ontology must create subsets. This does not work as a mode of reasoning of course, even if you work using retroaction, yet the breakdown here eventually becomes a constructive rupture. It transpires that, contrary to what Badiou asserts here, there are two 'axiom' systems or at least there are to some degree. Badiou is unable to account for states qua states in *Being and Event* in the end, just as he is unable to prove events or account fully for subjects. It is only in

Logics of Worlds that subsets are fully accounted for using the axioms of category theory, so there are two axiom systems. That said, category theory is not strictly-speaking an axiomatic system, it proceeds due to logical theorems on the whole, and in any case it is entirely dependent on the axioms of set theory for its logical theorems to work. All the same it will be the case that belonging is fully presentable in terms of set theory, but inclusion is not.

Badiou's conclusion is political and although not yet deserved, it will turn out to be true two decades hence. He says that inclusion has to be entirely dependent on belonging for the reasons given. This means that anything the state tries to do alone, so to speak, for example include terms whose multiple being it denies, will always be superseded by being as such or belonging. In particular, if one considers a state as an anti-void function, it wishes to deny the existence of the void by 'counting' or naming it, negating its non-relational in-difference and its content neutral indifference, this function will always fail. 'In particular, not only is it possible that the fixation of the void occur somewhere within the parts, but it is inevitable. The void is, necessarily, the ontological apparatus, the subset par excellence' (BE 101). Or, for a state to exist subsets must exist and if subsets exist then in every state there has to be a void. What this results in is a phrase of real philosophical importance: 'The integral realization, on the part of ontology, of the non-being of the one leads to the inexistence of the state of the situation that it is; thereby infecting inclusion with the void, after having already subject belonging to having to weave with the void alone. The unpresentable void herein sutures the situation to the non-separation of its state' (BE 101). This assertion could have political significance as well. A state is due to a void. In that a state is primarily an anti-void situation, all states are founded on the seeds of their dissolution. A state begins to dismantle itself by proving that in admitting to excrescence, which is ontologically impossible, it is secondary to ontology, and as ontology is woven from the void alone, states must admit to voids. In that their existence is predicated on being anti-void, all states must eventually fail.

But more significant, for us at least, is the philosophical implication which is, the very moment when a state which exists as relational, separated, quality-possessive and thus differentiated defining the existence of an excrescence, it admits that it is sutured to the non-separation of its state due to the void. In that this void is in-different, and that what it creates are quality-neutral, indifferent multiples, excrescences in a state negate their anti-void strategies due to the admission of indifference at the precise moment of their most profound differentiation, that between normal and excrescence parts. So, to differentiate in a state, a situation or a world, indifference is an essential component. In that states or worlds are defined as differentiated, relational assemblages or sets, they must differentiate to exist so they are reliant on indifference to exist. Finally, if this has political implications, are we not justified in saying that this not a modification of Marxist materialism, as most have said, but a different entity entirely, something that does not create a politics out of dialectical and material differences, but rather presents a politics of indifference? It is a question that becomes immediately pertinent in the ninth meditation when Badiou considers the relation of a state to indifference.

States and indifference (Meditation Nine)

Meditation Nine picks up on where Eight leaves off with a more pronounced political set of examples of the nature of the state of the situation. Central to these discussions are two issues we have already noted. The first is the classificatory nature of the state, which we said precludes any sense of a state of indifference. The second, related to this in some sense, is the fact that a state does not recognize individuals. Even when it appears to refer to an individual, 'it is always according to a principle of counting which does not concern the individuals as such' (BE 105). This leads to Badiou's strongly anti-indifferential argument that, because the state is always tied to an initially presented situation it cannot be indifferent. In his words, 'The state, solely capable of re-presentation, cannot bring forth a null-multiple – null-term – whose components or elements would be absent from the situation' (BE 106). He goes on to clarify this in terms of the example of a voter. When an individual, through voting, is represented by the state, it cannot be by virtue of being a presented term or as Badiou says 'as that multiple which has received the one in the structuring immediacy of the situation' (BE 107). Instead, 'this individual is considered *as a subset* . . . as the singleton of him or herself. Not as Antoine Dombasle – the proper name of an infinite multiple – but as {Antoine Dombasle}, an indifferent figure of unicity, constituted by the forming-into-one of the name' (BE 107). Still using the language of indifference Badiou goes on to elucidate that the state representation of subjects is always victim of a basic coercion that can go as far as resulting in death but is primarily Althusserian in its constant interpellation of its subjects. 'This coercion consists in not being held to be someone who belongs to society, but as someone who is *included* within society. The state is fundamentally indifferent to belonging yet it is constantly concerned with inclusion' (BE 107).

This muddies the waters a little. We have said that a state cannot be indifferent. This is true. It cannot admit to the void nor, for that matter, the event, both of which are indifferent; we have proved the former and will come to do so as regards the latter. In this instance, then, the voter is reduced from multiple potential, any multiple can potentially belong to any situation as they are totally indifferent as regards quality so they cannot be barred from a gathering because they lack the right connections or possess the wrong accent, to being assigned an identity such that they are forced into a subset grouping determined by qualities which are indifferent to who they are in their actual singularity. Here then we must differentiate singularity, the multiple that you are, from singleton, the means by which you are defined as that which is included in this subgrouping: {singleton}.

We pause here because in the history of the term indifference it is this kind of indifference that has regularly been demonized. This is the indifference of the state to the actual lives of its subjects through its rage to include irrespective of belonging. Yet we can see that this kind of indifference is facilitated by the differential nature of the state. There can be no subsets until the indifferent elements of the set are differentiated by being included in various subsets. If the state is indifferent, it is the cruel indifference of absolute differentiation not the profound ontological indifference of the void. And it is against this indifference of singular specificity that an event can occur. The rage to classify so as to negate the void is the very facilitating medium through which true singular events can be perceived.

The remainder of the chapter goes on, against orthodox Marxism, to discuss the means by which the state is determined not by binding, bringing elements into subgroups, but unbinding, doing this to resist its clear dependency on the void, in particular how the state does not define politics and is indeed not political. We will not linger here but we can leave these two provocative, related meditations with an axiom of our own. In being indifferent to elemental neutrality, the state is able to impose upon us identities of difference. As long as we continue to agree to occupy this paradox of identity formed from difference, as Badiou says, such a state is non-political, as Rancière might say, 'insofar as it cannot change, save hands, and it is well known that there is little strategic significance in such a change' (BE 110).[18]

We have now determined the nature of the one as pure multiple, and as the count-as and counted-into-one of the multiple as part of a set. In the first case we saw how the void is able to make multiples totally consistent in terms of belonging. In the second that however in terms of inclusion this consistency of all sets in general threatens the consistency of all the sets of this particular set or situation. This requires the metastructural re-count of elements that are included over those which belong. Yet, in that every state is a multiple of multiples, it still participates in excess and until Badiou can absolutely determine ontological stability from the smallest element to the largest he cannot propose a credible theory of the event. The final consideration of being, therefore, will establish the absolute stability of everything that is in terms of the nature and its relation to absolute infinity.

4

Nature and Infinity

The final word on being, before we move to the event, is that there exists an actually infinite nature which is absolutely stable. This statement brings together the three most important philosophical implications for set theory. The first pertains to the one, which is-not, and the multiple as the count-as-one. The second is the use of the void as a halting point or lowest level of the count, which then founds the stability of all other multiples. The third and final part is that if there exists a stability with a lower level which is not one, cannot be described as a single whole, then what is the basis of this stability? The answer is that this total stability, what Badiou calls nature, is actually infinite. We are able to define the stability of everything that 'is' as being a multiple of multiples due to the void, because we can show, mathematically, that actual infinities exist in a totally consistent, indeed tediously stable, manner. In practical terms, this reveals that Badiou's ontology completely depends on Cantor's proof that actual infinities exist. If this is proven, then a consistent stability exists beyond anything yet conceived of by our philosophy. It is this totally stable consistency, the actual infinity of nature composed of multiples of multiples but never counted as a total one, which is the only basis on which a truly inconsistent element, an event, can be thought. To get to this point we need to look at infinity from both sides now. First in terms of its actual existence as 'nature'. Then as regards the internal composition of all infinite sets due to the existence of actual infinities, in particular the questions of limit and succession or the exterior and interior of an uncountable yet totally consistent infinity.

Nature is normal (Meditation Eleven)

With the combination of the count of multiples which belong and the recount of submultiples which are included, a state of a situation can be said to be superstable at the metastructural level. Badiou is at pains to name this superstability nature and to demonstrate that it is, in his terms, 'normal'. To achieve this, he translates set theory into the language of the wider philosophical tradition. First through a powerful reading of Spinoza in Meditation Ten, then through a retranslation of certain well-known Heideggerian formulations into the language of set theory.[1] If nature is that which resides with itself or the 'remaining there of the stable' for Heidegger after the Greeks, for Badiou 'we will say that a pure multiple is "natural" if it attests, in its form-multiple itself, a particular con-sistency, a specific manner of holding-together. A natural multiple is a superior form of the internal cohesion of the multiple' (BE 127). A natural stability is, therefore, the re-count of the count or the making metastructural the basic structure of any situation of belonging. It is, by the way, worth bearing in mind that nature here is not an empirical concept available to the hard sciences and Badiou is not dismissing this view on nature per se. Rather, he is speaking of what counts as norm or normal in abstract modes of thought such as mathematics and ontology and how this normality in maths accords to the concept of nature in philosophy.[2] He goes on, reminding us of his three part typology of being as normal, singular and excrescent, and how in this instance normal being 'balances presentation (belonging) and representation (inclusion), and which symmetrizes structure (what is presented in presentation) and metastructure (what is counted as one by the state of the situation)' (BE 127). Badiou is admiring of this definition of normality as equilibrium and asks if stability derives from the count-as-one, his contention, then something which is counted twice like this is simply very stable indeed. At which point he retranslates Heidegger's more obtuse yet suggestive language by saying: 'Normality, the maximum bond between belonging and inclusion . . . Nature is what is normal, the multiple re-secured by the state' (BE 127–8).

Badiou now closes the small crack in the door made possible by his own typology. If, he asks, a multiple is normal, it could still be internally contradicted for example by elements which are presented by a multiple but not re-presented, so called singular multiples. In that for him there are three types of being, normal, singular and excrescent, we must make a stipulation if we wish to retain a stable norm. A normal multiple, we must concede, is made up of normal multiplicities alone, that all the multiples that belong to it must also be normal, and all the multiples that make up these multiples are normal, and so on. As he says more formally: 'a situation is *natural* if all the term-multiples that it presents are normal and if, moreover, all the multiples presented by its term-multiples are also normal' (BE 128). If we then take nature as normal we arrive at a crucial and perplexing theorem: 'nature remains homogeneous *in dissemination*; what a natural multiple presents is natural, and so on. Nature never internally contradicts itself. It is self-homogeneous self-presentation' (BE 128). Nature's stability, contra Spinoza in Badiou's reading, is attained due to the excessive nature of its dissemination. Indeed, this is the only way left for ontology, to prove the stability of situations due to the excess of the void and, in terms of the excess of the proliferation implied by the definition of being not being due to the situation of everything there is, its being a multiple of multiples. Nature then comes to be defined not as the totality of everything that is, but the consistent homogeneity of everything that can be. If something is presented as natural, it is because it is something that can be presented as natural, it is a normal multiple. There are an infinite number of these, but their infinity is actual and totally consistent in the procedural sense. If one says nature is a self-homogenous self-presentation, what one is really asserting is that nature is the actual and thus formally consistent infinity of the multiplicity of multiples with a clear halting point and no exterior upper limit.

Transitive sets: cardinal and ordinal (Meditation Twelve)

Meditation Twelve is concerned with a detailed consideration of the normality of the natural situation from a different direction, concerned as it is not with the language of philosophy but that of mathematics. Specifically, Badiou presents nature in terms of six basic presuppositions of set theory: normality as transitive sets, natural multiples as ordinals, minimality, the unique atom, nominalism and finally the proposition 'nature does not exist'. The meditation is brief, technical yet also important. We begin with transitivity. If you recall we mentioned in passing that normal multiples are ordinals not cardinals and that set theory is based on the acceptance of there being two kinds of number, ordinal and cardinal, or the difference between the number of elements in a set (countable ordinality) and its overall size (cardinality).[3] In finite sets, the size of a set is defined by the number of elements countable therein so that cardinal and ordinal coincide completely. In infinite sets, however, this is not the case hence the division established between the two kinds of number. When one considers the difference between having a number (cardinal) and being counted (ordinal) at first glance both seem differentiated. Yet the distinction between the two modes is facilitated first by the indifference of the cardinal number. A cardinal number is the same as another cardinal number if there is a one-to-one correspondence of all its elements. This is what we mean by transitivity. Transitivity states that if a set has the cardinal number 2, then I count each element of that set as corresponding one-to-one in terms of another set and I confirm they are both 2. Transitivity of this order is indifferent. What the elements 'are' in each set is irrelevant. All that matters is that the two sets are the same size. When I compare two sets for measure, for cardinal size, the count I use is a transitive count; this element is the same as this and so on, with no remainder meaning they are the same size. This can be done practically, in rare circumstances when we don't know how many elements there are in each set, but primarily the initial count of size of the set is through the enumeration of its ordinals. Surely this count cannot be indifferent as you need to know exactly how many elements there are by counting them, and you cannot count an infinite number in terms of ordinality, which for centuries was the incorrect sense of the infinite as a number too large to count? Yet Badiou shows that even counting is indifferent at root when it comes to transitivity.

Let us recall the set made up of no elements or the void set. Remember the void set and the singleton of the void set is not the same thing, by virtue of the separating law of collection. So, the void set is composed of no elements whatsoever, whereas the singleton of the void set counts as one the void, placing it in brackets. If we gather together these elements, the empty set and the singleton of the empty set, we end up with the set {Ø, {Ø}}. While Ø does not belong to the set, the void never belongs, it can now be included in the set in the second count. This count primarily allows us to count the empty set as {Ø} yet as every set includes the void so the set of the empty set is composed of two elements, the void set as counted and the actual void. This second set is fully transitive to the first. Obviously, it contains the void or empty set, in the second count represented as the singleton, {Ø}, but it also includes the void as such because it cannot exclude it. This set does not belong to the first count, but everything that does belong to that first counted set, which is nothing named as the void and presented as {Ø}, now belongs to the second set, counted as singleton element of nothing. So, although the second set has two elements and the first had one, the second is entirely transitive with the first due to the peculiar relationship of the void to belonging (it cannot) and inclusion (it is always included as the not belonging). The two sets are different sizes, but in terms of their elements they are transitive as each element has a one-to-one relation with each other in that they are all different presentations of the void, Ø, in terms of degrees of separation, {Ø}.

In addition, we can now say that the elements of the set {Ø, {Ø}} which we have shown to be transitive, are also in themselves transitive. It is obvious that

{Ø} is transitive in that what belongs to {Ø} is uniquely the void. The void is always included so what belongs to the set is also included. As for Ø itself, as it has no elements but is always included it cannot *not* be transitive as its only quality is that it is always included in any set. Transitivity is a set, all of whose elements are also included. A non-transitive set always has one element that is not a part of it in terms of one-to-one correspondence. As Ø has no elements, then it cannot be non-transitive and so, classically speaking, it is transitive.

Badiou so far has been counting ordinals. Now he gives our set, {Ø, {Ø}}, a cardinality: it is the number 2. It clearly has two elements. For the record the set Ø, {Ø} is the number 1, you could just write it {Ø}. What this means is that there are at least three transitive sets: 0, 1 and 2. Amazingly, from these three numbers all other ordinals can be generated using what are called von Neumann numbers. To count, basically, all you need are the three numbers 0, 1, 2 and the law of transitivity. You need 0 as your stopping point and as the generator of transitivity (there are no transitive numbers without the void as included nonbelonging). You need 1 as your base number, and you need 2 as the pair of your base number, or your recount. Each time you count, you tally a transitivity between 1 and 2 by virtue of 0 that, in recounting the 1 as 2, produces the next highest number. All numbers can be generated in this fashion and every element of an ordinal is an ordinal because they are all traceable back, due to transitivity, to this initial set Ø, {Ø} or the number 1 due to the void as not belonging yet includable.

Nature and minimality

The first parts of Meditation Twelve cover the transitivity of ordinal numbers in a compressed version of the discussion we have just had. Ultimately, we are led back to the point of the separation between ordinal and cardinal numbers we began with at which point Badiou says something else about the indifferent nature of the ordinal: 'One of the important characteristics of ordinals is that their definition is intrinsic, or structural. If you say that a multiple is an ordinal – a transitive set of transitive sets – this is an *absolute* determination, indifferent to the situation in which the multiple is presented'. From this he is forced to conclude 'Nature belongs to itself' (BE 134). This is, effectively and unusually for Badiou, an intensionalist argument with indifference used differently to refer to the mathematical specificity of nature as a self-generating system. Already, in that we know that no set belongs to itself and that Badiou thinks the one is-not so nature cannot be the count of all counts, we are prepared for the sixth part of the meditation, 'Nature Does Not Exist' (BE 140). For now, however, we can register a rare moment where Badiou appears to use indifference negatively. Because nature is self-generating, he suggests, it is always indifferent to that which is truly different: namely the event. An intricated system of total stability such as Badiou defines here as nature is one that remains 'emotionally' indifferent to that which is truly indifferent: the void and the event. This indifference of nature to anything indifferent, anything that cannot be differentiated according to the transitive sequence of ordinals 0, 1, 2, will turn out to be a decisive feature of nature in relation to the possibility of the event. Yet at the same time we can note the irony here. Nature, as a totally differentiated system of transitive, sequential ordinals with no gaps, no remainder and no exterior due to the differentiating relational nature of the second count, remains fundamentally indifferent. All numbers are reducible to the sequence 0, 1, 2 and, as we saw, 0, 1, 2 are not actual numbers but names for formally and quality neutral abstract operations: counts as one due to the void, counts as two due to the one.

The next section of the meditation moves towards an important concept which is the 'universal intrication . . . of ordinals' or the thesis 'There are no holes in nature' (BE 136). This marks the end point of what Badiou calls 'the organic concepts of natural-being'. These four concepts are normality as transitivity, order, minimality and finally total connection. Together they define an unassailable stability of nature or of everything that can be said to be. To get to total connection, which is our goal and which we have yet to consider, we must then look again at order and minimality. As regards order, we saw that sequentiality within ordinals presents the order of the natural such that it has no holes. Sequence is an important concept especially going forward into *Logics of Worlds* and here it is defined in terms of being smaller than or $<$ in a manner that bisects with the later volume. In a logical structure, structural ordering can be vouchsafed due to three comparators, larger than, smaller than and equal to, reduced to a single symbol $\leq$, in category theory at least. As regards natural ordinals, however, we need only smaller than, $<$, because, as we saw, counting ordinals depends on the repetition of already established numbers and so is always descending.[4] In addition to sequence, the law of belonging means that 'if one "descends" within natural presentation, one remains within such presentation' (BE 134). This allows Badiou to explain precisely what smaller-than means in this context.

Defining smaller-than is problematic philosophically as the issue of measure has haunted ontology since its inception. Determining the limits of an object to such a degree as it is measurable against another object has proved all but impossible. Yet again, however, we see that if we consider measure indifferently in terms of abstract transitivity, this problem dutifully withers on the vine. As Badiou says, due to the law of transitivity, 'That an ordinal be smaller than another ordinal means indifferently that it either belongs to the larger, or is included in the larger' (BE 134). This indifferent measure has nothing to do with actual size, whatever that means, and rather means something closer to directly subordinate to which is what is meant by being well-ordered in set theory. Indifferent measure then concerns merely the property of relation, belongs or does not belong, is included but does not belong and so on, allowing a degree of comparison that is determined entirely by relationality; the true meaning of smaller-than. Again there is a robust lesson in the power of indifference here in that we are able to compare two elements in terms of 'size' without knowing anything else about them except their comparative relation as regards the elements they contain. This means naturally that no set can be smaller than itself because size, smaller-than here, is determined by comparison of belonging and no set in nature belongs to itself. It also means there is a smallest element. This is determined simply by saying if a multiple exists with a certain property, as regards the relation of order, there is no multiple smaller than this first multiple that possesses this property. Due to the laws of sequence, belonging and transitivity, this way of determining the smallest will always function and, as we saw, the whole string of ordered ordinals also has its smallest element which is 0.

Badiou is well aware of the ontological import of such a proposition. Speaking of the property of miminality he says:

> What it does is orientate thought towards a natural 'atomism' in the wider sense: if a property is attested for at least one natural multiple, then there will always exist an *ultimate* natural element within this property. For every property which is discernible amongst multiples, nature proposes to us a halting point, beneath which nothing natural may be subsumed under the property (BE 135).

Minimality due to indifferent descending comparative relation, $<$, solves one of the two classic problems of metaphysics and ontology. In determining the thing as the thing it is due to the properties it possesses, at what point do you leave off parsing said properties to avoid infinite regress: this is the property of X which is the property of X1 which is the property of X2 ...? If you indifferentiate the element in question, which one can only effectively do through set theory axioms, then your mode of comparison is of a different order. Instead of comparing due to properties possessed, you are comparing due to elements which belong. It is a comparison of pure relation not of possession and, as we saw, whenever we move from possession to relation we are able to indifferentiate the object in terms of qualities, yet still retain the right to determine the being of said object. And of course again due to the indifferent nature of set theory, this time its dependency on the in-difference of the void, there is a halting point for this. So we can always determine either a minimality of relation or a foundational limit to relation as such when speaking of the 'size' of a set in nature, size being the only property it possesses which, in any case, is not a property as such but a relation which it does not possess per se.

Nature and intrication

Order due to minimality solves half of the metaphysical puzzle, so to speak. Yet it also provides us with another astonishing quality as regards nature defined in this strictly ordinal, relational and indifferent fashion. Badiou reveals now that 'The principle of minimality leads us to the theme of the *general connection* of all natural multiples. For the first time we thus meet a *global* ontological determination; one which says that every natural multiple is connected to every other natural multiple by presentation. There are no holes in nature' (BE 136). This may not seem obvious as the strict order of minimalization actually seems to preclude the more Deleuzian or Nancyean concepts of reticulation that we tend to think of when we consider global connection. If Deleuze favours the rhizome over the arboreal, Badiou's connection resembles a branchless trunk firmly rooted but with no upper limit; a kind of ontological rope trick. Yet again this is perhaps what is so radical about his proposition. Not that everything is connected, but the arid nature of said connection.

As he says, due to the law of transitivity, belonging, sequence and minimality, by definition 'Every ordinal is a "portion" of another ordinal . . . insofar as every multiple whose count-as-one is guaranteed by an ordinal is itself an ordinal' (BE 136). Natural order then resembles something like a modern terrorist cell in that one element is connected to the next element and so on down the line but the third element in the chain, for example, never actually meets the first, nor will it meet the fifth. Not only is this natural state one of relational ignorance, you know nothing other than your two closest neighbours, it is also, paradoxically, claustrophobic, due to the law of universal intrication:

> Because every ordinal is 'bound' to every other ordinal by belonging, it is necessary to think that the multiple-being presents nothing *separable* within natural situations. Everything that is presented, by way of the multiple, in such a situation, is either contained within the presentation of other multiples, or

> contains them within its own presentation. This major ontological principle can be stated as follows: Nature does not know any independence ... the natural world requires each term to inscribe the others, or to be inscribed by them. Nature is thus universally *connected*; it is an assemblage of multiples intricated within each other, without a separating void ... (BE 136).

The inexistence of nature

The final section of this truly amazing meditation is the punchline of the whole consideration: "Nature does not exist". Following the logic of ordinals Badiou reveals: 'If it is clear that a natural being is that which possesses ... an ordinal, what then is *Nature* ...? nature should be natural-being-in-totality; that is, the multiple which is composed of *all* the ordinals' (BE 140). He then goes on to show that for nature to be all the ordinals it must itself be an ordinal and so be transitive, otherwise it could not collect all the other ordinals. Collection needs belonging and inclusion due to sequential ordering in terms of the operator $<$. This being the case, through the application of simple logic which the reader can pursue for themselves if they wish, 'Being an ordinal ... the supposed set of all ordinals must belong to itself ... Yet auto-belonging is forbidden' (BE 140). We saw this coming of course. If one grounds nature, one cannot top it off. Badiou is not perturbed by the various disturbing discoveries as regards nature not least its inexistence. He simply says blithely: 'Nature has no sayable being. There are only *some* natural beings' (BE 140). In other words, natural beings exist, but a totality of all natural beings called nature does not. This is another way of saying being is-not. There is no total multiple of the multiple of multiples. Every multiple must be a multiple of multiples or it doesn't exist.

Badiou closes this section with the wonderful formula of natural multiplicities that results 'on the one hand, in the recognition of their universal intrication, and on the other hand, in the inexistence of their Whole. One could say: everything (which is natural) is (belongs) in everything, save that there is no everything' (BE 140–1). A high risk strategy because it seems to neglect the basic ontological question: Why is there something rather than nothing? Which makes Badiou's operational solution all the more brilliant: 'The homogeneity of the ontological schema of natural presentation is realized in the unlimited opening of a chain of name-numbers, such that each is composed of all those which precede it' (BE 141). All because of that initial proof that the one is-not, we have the natural universe of ones, thanks to the sequence of three ordinal 'numbers' and their defined normality due to transitivity, order, minimality and total connection, it's as easy as 0,1,2.

Potential and actual infinity

The total stability of nature, along with its inexistence, depend entirely on the one quality of nature that we have not yet full discussed, its being an *actual* infinity. What does this mean in practice? Since Aristotle, there has been a clear division between actual and potential infinities. An actual infinity is the totality of all things, the universe, and Tiles asks us to think of this as something absolutely complete. Taken as complete, this actual infinity has no limit because it composes every thing and thus there is nothing outside of it from which to judge its completeness. Notably, we could add that a complete infinity of this order is in-different in that it cannot be differentiated from some other thing. There is no other thing. A potential infinite is of another order. It is just the uncountable set of numbers in a series, famously how many points on a line. In terms of the points on a line and Zeno's paradoxes, the issue is one of infinite divisibility.

Aristotle comes down on the side of potential infinites, there are certainly many examples of infinite divisibility and uncountable series, and rejects any theory of actual infinity. For Tiles, this prejudices the idea of the infinite ever since as potential infinity, that which lacks completeness. Since the Romantic age, philosophy and Christianity came together in finding a use for the 'actual' infinite and that use was God. Even Cantor invokes God to resolve the irresolvable elements of his set theory, in particular, the imposition of God as an absolute limit which serves to separate Being from the world of beings. This is the famous difference in Hegel between bad infinity, that of infinite divisibility and proliferation in mathematics, and good infinity, the imposition of an absolute limit of totality: completeness. After Cantor, an actual infinite is proven to exist to the satisfaction of the majority of mathematicians. In effect, Cantor proves you can count the uncountable. This is achieved by splitting the number into two, ordinal and cardinal, changing the way we count, and using axiomatic set theory to retroactively prove the existence of the infinite.

Cantor's actions negate the artificial division into potential and actual infinity. As far as modern mathematics is concerned, potential infinity is actual infinity. In this light, actual infinity, completeness as totality, is impossible, there can never be a one of all ones, but this does not matter as it is no longer needed.

Instead, due to actual infinity as the counting of the unlimited series, an absolute stability without One-ness exists. This is what Badiou terms nature.[5] In a basic sense the proof that actual infinities do exist, in other words that we can count potential infinities putting a limit of sorts on the unlimited, solves the problem of the one through that of the stability of the many. If we take the local 'one', divest it of qualities and internal division, and simply define it as a unit that is collected with other units whose being is entirely relational not quality possessive, extensional not intensional, then the division of the problem of the one and the many disappears. There is now no one, no total One, and so we do not have to think of the many as parts of such a One.

This literal revolution in thought is facilitated by Cantor's great innovation in thinking part and whole due to the logic of sets and subsets. Recall that due to the axiom of excess or the powerset there are always more parts to a set, more subsets, than there are counted initially as the whole of that set. Cantorian set theory was able to show that parts were not smaller than wholes, but of the same 'size'. This observation negates the part-whole dialectic hierarchy and in so doing puts an end to millennia of rather pointless reasoning. Finally, we could add that the nature of the one we have just described is facilitated by its being indifferent, and said indifference enables the one-to-one correspondence of parts to whole as regards infinite sets. This would be a fitting conclusion for our consideration of infinity if it were not for the emphasis Badiou places on the importance of actual infinity and the nature of his explanation of Cantor's 'proof' that actual infinity exists. We need to go further.

Proving the actual infinite

There are three central elements that allow Cantor to state that actual infinities exist. These are recursion, acceleration and bijection. In particular, these three qualities allow one to count in such a way that serial infinities are 'countable'. This also requires that we divide the number, as said, into two elements or two ways of counting: ordinal rank and cardinal size. Ordinality is a recursive relation. Classic examples of recursion given by the literature are descent – if a child is a descendant then a child of a child is a descendant and so on – and stroke counting, or a sequence of written strokes. Natural numbers, as we saw, are also defined in this way. Recursive counting means that you can count locally without consideration of the global to such a degree as you can 'count' all the natural numbers, the usual stumbling block for theories of infinity, which we will call here set N. If you accept zero is the initial natural number, and the successor rule that for every X, if X is in N, then the successor of X is in N, you can define the total set of natural numbers. For the record here is the proof: if 0 is in N and, for all X, if X is in N then X + 1 is in N. To this we add the proviso, and there are no other objects in N. These three rules, Steinhart calls them the initial rule, the successor rule and the limit rule, compose some of Cantor's greatest formal discoveries.[6] Cantor's three rules capture the entire process of counting to the infinite in three simple axioms. If there exists the initial number 0, Cantor says, then for every number n, there exists a successor number n + 1. This will generate all the positive finite numbers in terms of their ordinal rank, not in terms of how many parts there are to the set (cardinal size). Finally, Cantor applies the limit ordinal rule that for any endlessly increasingly set of numbers there exists a limit number greater than every number.

The limit ordinal or 'second' existential seal, the first being the void or nondivisible smallest point, is crucial and we will return to it in detail especially as Badiou sees it as an exceptionally important point. For now, satisfy yourself with an easy shorthand of the limit rule as second existential seal. The limit or seal is the proviso that there are no other objects in N. This gives us the set of all the objects N. This is possible as set theory defines the number of objects in a set due to the collectability. Thus elements in the set *are* the set. As we said normally we would say this number was the cardinal, the measure of the set, but due to recursive reasoning we can say that if the set is all the elements which belong to it, we can count each element through its relation to the previous element. This relation means we will never count anything other than these kinds of numbers, and that we can count 'all' these numbers without actually 'naming' any number other than zero. As we saw due to van Neumann numbers all sets are made up of counting zero effectively. Another way is to talk of what we call the union in set theory. The axiom of union is one of the nine axioms of ZF+C we have not yet considered. The union is the combination of all the elements in two sets. So, the limit rule could also be called the union of all the elements in recursive relation to the natural numbers. Note we haven't actually counted the natural numbers. Instead, we have used abstract notation and axioms to prove the existence of a set of all natural numbers. This set is the first actual infinity and was called, by Cantor, ω.

Von Neumann's ordinals are another way of clarifying that at least one actual infinite number exists: ω. Every number, for von Neumann, is the set of all numbers less than it: ordinality. So, if we start with 0, n is all the numbers less than zero. There aren't any so we have an empty set $0 = \{\ \}$ or the number 1. Then as regards succession, for every number n there is $n + 1$. As every number is a set of the lesser numbers, $n + 1 = \{0 \ldots n\}$. Now to the limit or seal. The number ω is the set of all numbers less than ω. Due to the initial and successor rule, every finite number is less than ω. As all finite numbers are less than ω, so it is that $\omega = \{0,1,2,3 \ldots\}$. As these rules generate numbers in a linear localized sequence, by definition ω is in sequence as the sequential end to the sequence. This permits us to say that ω is an ordinal number defined simply as the last in the sequence of all natural numbers. We don't have to count ω globally as recursion allows us to do it locally. Hence in splitting the number into ordinal and cardinal, and using the three rules of successive counting Cantor gives us, with clarification from von Neumann, we have proven the existence of an actual infinite number, the set of all natural numbers.

Doubling and Dedekind infinites

Another way of defining the infinite set is to use Dedekind's definition of the infinite set as that in which the parts are equinumerous with the whole. This conception of equinumerosity will now allow us to begin to think of the possibility of there being more than one infinite number and more than one type of infinite number. Dedekind's famous definition of infinity is refashioned in Tiles as 'an infinite set can be put into a one–one correspondence with a proper part of itself, no finite set can' (Tiles 62). By proper part is meant a part all of whose parts belong to the set. Why does this result in infinity? If a set can be put into a one-to-one correspondence with a proper part of itself, it means that the interior of the set, the parts it contains, will always be greater than the total 'size' of the set or its cardinality. The view that the parts are greater than the whole, or that the many exceeds the one is not news to our tradition but Dedekind and Cantor's concept of an infinite set, where the parts of the set are larger than the whole, is much more well-ordered that that found in continental philosophy. For example, although the set of natural numbers is smaller than the elements contained within it, there is still a set of natural numbers, ω, the first proven, actual infinite set. For this set, the fact of there being more parts than the set can enumerate does not undermine the set. Instead, and this is what makes Dedekind and Cantor truly modern according to Badiou, the definition of an infinite set as one whose parts exceed the whole is what secures the actuality of their being an infinite set and thus that there is a stable natural state of total transitivity. To understand this, we have to pay attention to the idea of one-to-one correspondence or equinumerosity.

Tiles says 'When two sets (whether finite or infinite) can be put into a one-one correspondence with each other they are then said to have the same power or cardinal number' (Tiles 63). In the case of finite sets, as we saw, the cardinal number of a set is the same as the ordinal, but in infinite sets this is not the case: 'in the infinite case it will mean that an infinite set may have the same cardinality as the proper part of itself' (Tiles 63). To prove this Cantor uses the process of bijection, for example in the case of one-to-one correspondence between natural numbers and even numbers. You can establish a one-to-one correspondence with the two by setting up the following table:

$$0\ 1\ 2\ 3\ 4\ 5 \ldots n$$

$$0\ 2\ 4\ 6\ 8\ 10 \ldots 2n$$

Let us go now to Steinhart's technical definition of bijection in relation to infinite sets. Speaking of the recursive method he notes: 'A recursive definition is a finite way to describe a set whose cardinality (whose size) is greater than any finite number' (Steinhart 156). The wording of this definition is very careful. First, recursion is finite; it is a limited set of rules.[7] Second, these rules don't count the cardinality of the set but describe it. This is in accord with Tiles' point that for the parts of a set to be greater than the whole contravening one of Euclid's most central and lasting axioms, 'It seems that the part–whole relation and notions of magnitude based on it have to part company with the application of numbers' (Tiles 63). Infinite sets are without number, hence infinite, because they cannot be counted, she says, 'but also because even the notion of cardinal number, as a measure of size or numerosity, can get no grip here' (Tiles 63). Back with Steinhart, clearly recursion allows us to describe a set that we cannot count and it is this point the reader needs to dwell on. All of Badiou is contained therein and all of indifference.

Axiomatic set theory allows one to describe formally a situation that cannot itself be counted or presented inductively – empirically. So that we can show how there must be infinite sets using basic operations that can be proven, without ever being able to see what we are describing. As Steinhart says: 'A set S is *infinite* if there exists a proper subset T of S such that the cardinality of T equals the cardinality of S. Equivalently, there is a bijection (a 1–1 correspondence) from S onto T. A set that satisfies this definition is also known as a *Dedekind infinite*' (Steinhart 157). Unpacking this for the non-specialist, if we take the set of natural numbers N, and the subset of even numbers E, as every even number is a natural number, E is a subset of N; a 'proper' subset as there are also numbers which are not even which are subsets of N. Bijection comes about by simply doubling each number 1 = 2, 2 = 4, and so on. 'Doubling associates each number in N with an even number in E. And the inverse associates every even number in E with a number in N. So doubling is bijection. Hence there are exactly as many even numbers as numbers' (Steinhart 157). Hallward describes this perhaps more clearly when he says: 'Since the sequence of natural or "counting" numbers is itself infinite, it is perfectly true that there are as many odd numbers as there are both odd and even numbers'.[8]

Centrally, for our purposes, Steinhart presents his definition of a finite set on the basis of the infinite set: 'A set S is *finite* if it is not infinite' (Steinhart 157), and this was indeed one of the most incredible results of the Dedekind infinite. Up to this point we had defined the infinite as that which was without limits, thus the finite came first and was stable basis for the definition of the instability of the concept of infinity. In contrast, now we define the finite as that which is not infinite because, in fact, there are an infinite number of infinite numbers and an infinite variety. Infinity is, therefore, the stable norm and the finite is actually very rare and rather special in comparison.

Equinumerosity allows us to prove the existence of an actual infinity but we can go further and prove, quite simply, that there is an infinity of actual infinite sets. Using von Neumann's definition of numbers, after we have proven the first infinite set, the set of all natural numbers, using the logic of recursion, we can now add another number to ω or the set of infinite numbers. If infinity was produced from $n + 1$, then an infinity of infinities can be produced with $\omega + 1$, $\omega + 2$ and so on right up to $\omega + \omega$ or ω^2. This then can also be expanded using the same logic leading eventually to ω^ω which is the multiplicative series. This is not the end of the series. Having proven that there is at least one infinite number, you can show that there is an infinite series of infinite series. As we saw, in finite numbers the ordinality and cardinality of said sets is always the same. We now find out that whatever the specific ordinality of an infinite number, the size or the cardinality of all these infinities remains the same. We don't need to worry about the proof of this except to say that because finite numbers cannot be put into one-to-one correspondence with the infinite, otherwise the finite would be infinite and the infinite would be finite, the first smallest number that is the same size as the first infinite ordinal must be ω itself because that is one of the laws of cardinality: the smallest number that can be put in 1-1 bijection is the number itself. Now, because all the infinite ordinals can be put into 1-1 correspondence, that is what defines each of them as infinites, we have to say, using the same logic, that while the ordinality of these numbers is different, in each case the size, the cardinality, has to be the same. This seems to make sense. If you cannot delimit a number in terms of counting, then how can you differentiate infinite sets? Yet Cantor proved that not only did at least one infinite set exist, but in actual fact an infinite number of actual infinite sets exist, and, going even further, due to his 'diagonal' argument, there exist infinite numbers of different size or different cardinalities.[9] It is this discovery, that there are not only an infinite number of infinities but that they are of an infinite number of different 'sizes' that will allow Badiou, eventually, to prove that there is such a thing as the event.

We are now in a much stronger position to understand Badiou's complex relationship with infinity. We have seen that von Neumann ordinals, which are indifferent, allow for the generation of Dedekind infinities, due to the indifferent logic of bijection. We have seen how Cantor's basic three-part machine, zero, successor, limit, is able to generate with ease at least one actual, infinite number. We have also seen that there is a stable infinity of infinities, and finally that there are infinities of different sizes or cardinalities. We know a great deal about actual infinities but apart from the fact that there are actual infinities, an infinite number of them in fact, in terms of infinity itself we don't need to know much more than that. Actual infinity proves the stability of nature and is the final piece in a three-part ontological puzzle: there is the void, there is the count-as-one, and now there is an actual infinity. From this we can say with absolute certainty that due to the void or the subtractive nature of being, we can prove that all that exists are multiples of multiples. These multiples of multiples can, due to the indifferent difference of counting them as belonging and included, generate an infinite number of situations. These situations, due to the bifurcation of number into cardinals and ordinals, can be proven to be actually infinite; that is, uncountable in a stable, operationally transparent, indeed everyday fashion. So we can say that using an immanent system of count-as-one nature, all that there is, can be shown to be absolutely consistent by describing it as an actual infinity of actual infinities, without recourse to the One, and avoiding all the logical problems attendant on infinity from Aristotle to the nineteenth century. The discovery of actual infinities, then, is the single most important discovery of the modern age at least in terms of formal or abstract reasoning. It means simply that if the one is-not, consistent worlds can 'be' due to the fact that their actual infinity, thanks to a clearly defined halting point and a simple set of recursive modes of reasoning, is incontestable and ultimately banal. In effect, in terms of ontology, nothing more needs to be proven. However, for Badiou, the stability of nature as an infinity of infinities with no One or complete whole, is really only the start of the journey. Being is relevant, for him, only in terms of how it can produce an event. Yet what nature shows is that change of all kinds is part of nature. Difference is absolutely stable. There is no exterior other, and the interior other is entirely circumscribed by the laws of cardinal and ordinal numerical operations. If Badiou is going to be able to find an actual difference to go alongside his actual infinity it will need to be generated from novel means, in particular, as we shall see, from the interplay between limit and succession that makes up the rather unique finite as a result of the totally ubiquitous infinite.

The limit

For all the attention given to the innovation of the actual infinite, the shocking conclusion Badiou draws that the infinite is the precondition for the finite means infinity is relegated to the commonplace. The literal Romanticization of the infinite as the absolute limit of being was simply a poetic trope, no more. Infinite series, infinite sets, infinite multiples of multiples, even different kinds of infinities (an infinite number of kinds) exist everywhere because they *are the everywhere*: they are nature. These infinities are absolutely well-ordered due to their inherent transitiveness and bijective powers, a point that contravenes Hallward's doubt as to whether every situation is to be called infinite.[10] If it is a situation, it is part of a natural order, and as such it is infinite because it has no countable whole. While it is interesting to say such a natural order is infinite because it is indifferent, it is not particularly interesting to talk of natural order beyond that. For once we can say, in a negative register, that the infinite natural situation is indifferent in that it is totally without significant difference. Instead, Badiou explains, what becomes notably more interesting, more significant, within the indifferent order of infinite, transitive, well-ordered sets by virtue of the bijective nature of recursion, is that which we thought we had the measure of: the limit as such. In fact, Badiou spends less time considering the specificity of infinity than he does of the finite itself.

To better understand the revelation of the assertion, the limit ordinal exists, ω or the first actual infinite, and grasp its precise nature, we need to think again about Cantor's three rules used to generate the infinite set based on the logic of well-orderedness. We also need to spend some time amongst the pages of *Number and Numbers*, the follow-up to *Being and Event* without which, certain aspects of the main text remain obscure, certainly to the non-specialist. Badiou explains that a well-ordered set is one in which between each element there is a 'relation of *total order*'. This order, the essence of structure, is greater than, smaller than or equal to. In a well-ordered set 'no two elements are "non-comparable" by this relation' (NN 53). If you struggle with what non-relationality means for Badiou, here it is: to be without ordered rank.

To this simple definition you must add one more proviso: 'given a non-empty part of the set so ordered, there is a *smallest element* of this part (an element of this part that is smaller than all the others)' (NN 53). This is the axiom of foundation designed to negate infinite regression due to the peculiar qualities of the void.[11] One of the common errors when considering a well-ordered set is to think of the set as defined by its last element, the set of the set or its maximal element. In a normal, worldly situation this certainly exists, but note when it comes to infinite sets it does not. (We must say that eventually we will come to see that all worldly situations are actual infinities). Although we spoke of the third rule, the second existential seal, as defining the limit of an infinite set, the crucial definition of the limit of a set is not its upper limit, after which nothing succeeds, but its lower limit, that it succeeds from something else. We can see this when we realized that all Cantor did in effect was define the infinite set as well-ordered. If this had depended on putting an upper limit on the set this would not have been possible. Instead, he saw that the set is defined as being the successor to the upper limit of another set. This means that ω can be defined as infinite, limitless, as being the limit ordinal to the finite set of all natural numbers, just as natural numbers as a set are defined, through von Neumann's ordinals, as being the successor set of the limit ordinal 0.

Using well-orderedness Cantor is able to treat ω as the limit ordinal of the natural numbers, opening up the infinity of infinite sets we have described. To achieve this, he needs to overcome the problems of the smallest, of succession and the one. We have dealt with the one already so we are left with the (lower) limit and succession, yet we have also explained these have we not? Here then Badiou asks us to think again, but in a different register: 'we must distance ourselves from operational and serial manipulations ... The establishing of the correct distance between thought and countable manipulations is precisely what I call the ontologization of the concept of number ... we must abandon the idea of well-orderedness, and think ordination, ordinality, in an intrinsic fashion ... We do not want to count; we want to think the count' (NN 58). What is clear from this is that at no stage is Badiou 'doing' mathematics, or philosophy of mathematics.[12] Instead, he asks us to think mathematics from the perspective of philosophy. When we think the limit ordinal we don't need to prove it but to think what the limit ordinal actually means.

The background to this is the relation of well-orderedness to the Continuum Hypothesis (CH) which plagued Cantor throughout his life and in more romanticized versions of his biography eventually cost him his sanity. We will treat this hypothesis in detail when the time is right but at this stage it is illustrative to think of infinity as a philosophical challenge rather than a procedural result, Badiou's demand, by placing well-orderedness and CH together for a moment. CH is the name for Cantor's belief in a well-ordered relation between ordinal infinities, and the transfinite numbers of an infinite number of types of infinity or cardinal infinities. To recap, due to bijection and well-orderedness, Cantor was able to use numerical, recursive succession, von Neumann numbers, and one-to-one correspondence to show that real infinities exist even if they are 'uncountable'. The first infinite ordinal, ω, is in this sense well-ordered as is the next, $\omega+1$, and so on all the way up to an infinity of ordinal infinities. However, when it comes to cardinal infinities or the infinity of real numbers, Cantor wished to show that there was as direct succession from the first ordinal infinity, sometimes written ω_0, to $2^{\omega 0}$, the first cardinal infinity.[13] Cantor's 'diagonal proof' was able to show that an infinity of real numbers can be proven using one-to-one bijection, but he was unable to show that there were not other infinities between the two kinds of infinity. This meant that while he knew that different kinds of infinities existed, as there was no well-ordered successive relation between the two infinites, ω_0 and $2^{\omega 0}$, he was unable to say how big the cardinal infinities actually were. They became, at this stage, non-numerable or inconsistent numbers. Cantor found this unacceptable, Gödel proved CH was impossible, and Cohen used the axiom of choice to show that it didn't matter as inconsistent infinite cardinals or transfinite numbers could still be make to be procedurally consistent without recourse to well-ordered succession. Why this matters at this stage is that Badiou is about to rethink the concept of succession due to lower limit, the basis of well-orderedness, from a philosophical perspective to celebrate the radical possibilities of a gap between two successively ranked numbers. This will then open up the transfinite, inconsistent cardinal numbers as a basis for the inconsistent multiple he calls the event. He will achieve this by using Cohen's reconfiguration of the axiom of choice in determining the consistent order of a non-denumerable multiple, giving the event the status we already mentioned of an inconsistent multiple that is treated with procedural consistency, although not quite that of well-orderedness. Finally, in the second volume of our study we will see that well-orderedness is the basis of *Logics of Worlds* and its radical representation of the event, so everything Badiou says about ordered modes of successive recursion is of great importance. With this clarification in place, let's now think about succession and lower limit not in terms of producing order, but as a means of disrupting it.

Succession and limit

The ninth chapter of *Number and Numbers*, 'Succession and Limit. The infinite', takes its name from the second section of Meditation Fourteen and explains with greater clarity the technicalities of successive recurrence and limitation with which I believe the reader is now sufficiently familiar. Yet it is so much more than this in its courageous attempts to think the passage, the limit, on the basis of set theory, in an entirely new way for the future of philosophy and without its augmentations, a significant portion of *Being and Event*'s interaction with the infinite appears to be missing. Badiou starts by reminding us that to pass from n to n + 1, its successor, is not the same kind of passage as that from natural numbers to their beyond, ω. This allows him to say that his project is not to think about succession as such 'as an intrinsic quality of *that which* succeeds as opposed to that which does not,' but to seek the '*being of the successor*' (NN 74). The first element of this being is the means by which an element is added to a set. Using von Neumann's ordinal, we will call it W, in such a truly transitive set the set's elements are composed of all the elements of W and W itself. To everything that makes up W, we also add W itself. This is a new element, W is never an element of itself as no set can be, but this addition is not truly added either. Badiou says

> it is not a matter of an extrinsic addition, of an external 'plus', but of a sort of immanent torsion, which 'completes' the interior multiple of W with the count-for-one of that multiple ... The +1 consists here in extending the rule of the assembly of sets to what had heretofore been the principle of this assembly, that is, the unification of the set W, which is thereafter aligned with its own elements, counting *along with them* (NN 74).[14]

We have learnt our first lesson of successive counting. We do not really add one more to the set but rather we add the set to itself to produce a new set: the set of the elements of W. Thus, the set gets 'bigger' internally through the act of torsion which allows us to name the set W as W, adding a new element to it, that does not add something from the outside, but expands the set from the inside. The second element of counting as succession is also mentioned here, that this new extra element does not so much come after, and is not added from the outside, but comes from inside and runs alongside the other elements.

Badiou formalizes this by calling 'the set obtained by adjoining W itself, *as* an element, to W's elements' the successor ordinal. He then develops more this non-sequential and non-hierarchical view of counting which seems to fly in the face of what we have defined as well-ordered successor, larger than, smaller than or equal to. He says: 'The idea of the "passage" from two to three ... is, in truth, purely metaphorical. In fact, from the start *there are* figures of a multiple being, D and T, and what we have defined is *relation* whose sole purpose is to facilitate *for us* the intelligible passage through their existence .... We therefore think, in the succession T = S(D), a relation whose basis is, in truth, immanent ...' (NN 75). There is, it would appear, no actual succession. All these numbers already exist so that their successive definitions are merely an anthropological means for humans to be able to think these numbers. Badiou is at his most anti-constructivist and extensionalist here by insisting that numbers actually exist, whether we do or not. When we speak of T as being succeeded by D, what we actually say is that 'T has the structural property ... of being successor to D' (NN 75). Its being a successor is its very being, and when we define it as part of a successive sequence, this is an artificial causal dependency that does not truly exist but is merely constructed. Ordinals are ranked, or better rankable, but they all exist at the same level irrespective of rank. The simplest way to say this is that when an ordinal is defined in terms of succession, its Being becomes that it succeeds. In that it appears in a situation as a successor ordinal, it succeeds, but this does not negate the truth of the existence of the number nor does it delimit the existence of that number to only succeeding.

To sum up, an ordinal is defined ontologically, not numerically, as having the property of succeeding. One can easily call this indifferent succession and say that all ordinals are the same in that they succeed. This is important as if you want to speak of infinite ordinals, qualitatively different from finite ones, the same rules can indifferently apply: they are ordinals not due to their rank but because they possess the function of succeeding.

The upper or maximal limit

Moving on from succession, Badiou now considers the maximal element in rather similar terms. Starting again with the set W we can say it is composed of all the elements of w_1 and also w_1 itself. This is just another way of saying what we have just said. This being the case W is a successor of w_1. This then allows us to deduct a maximal element from succession. 'Let's agree to call the *maximal element* of an ordinal the element of that ordinal which is like w_1 for W: all the other elements of the ordinal belong to the maximal element ... *An ordinal will be called a successor if it possesses a maximal element*' (NN 76). This definition confirms the intrinsic and immanent means by which we can define a successor ordinal: 'The singular existence of an "internal" maximum, located solely through the examination of the multiple structure of the ordinal ... allows us to decide whether it is a successor or not'. This leads to 'an immanent, non-relational and non-serial concept of "what a successor is"' (NN 77). From this, Badiou then asks the question are there ordinals which are not successors and concludes yes, the void is an ordinal that does not succeed: 'the void is itself *on the edge of the void*, there is no way it could follow from being, of which it is the original point' (NN 77).

This is another way of confirming Badiou's nonrelationality, which we can compare to the nonrelation of the successor ordinal. The nonrelationality of the void is due to its inability to succeed, while the nonrelationality of every other ordinal is the means by which it can always be defined as succeeding internally without the need of any other number. As we have stated, any situation of nonrelationality is one we are taking as indifferent. The void is in-different because it does not succeed, nothing comes before it from which it can comparatively determine itself. Yet the very definition of the ordinal weakens or at least thickens our definition of in-difference here. Actually, ordinals are also nonrelational in that they can succeed from inside the same set, making the set bigger. We are learning then to think of differentiation in an entirely new fashion, one that will facilitate a view of the event as such as truly indifferent difference.

Returning to the text, having proven that there is at least one non-successor ordinal, Badiou asks if there are others. If there are, then he proposes that we call them limit ordinals or those ordinals which do not succeed. So, a limit ordinal, ω will be the first, is an ordinal greater than 0 that does not own the property of succeeding. Important in what follows is the realization that while limit ordinals do exist, we cannot prove them, for in fact their lack of proof means that we need to define them in a manner that is realist, decisionist, extensional and evental. We must assert that limit ordinals exist in an immanent and intrinsic fashion, because we cannot prove that they succeed as they themselves do not succeed a deductive, lower proof, basis of all numerical proofs as we have repeatedly shown. If limit ordinals exist we must say they exist in a non-successor fashion.

Badiou believes that if limit ordinals exist they must be structurally different from successor ordinals. He asserts, for example, that no ordinal can come in-between an ordinal (W) and its successor (S(W)). He goes on to prove this but I think the truth of this is self-evident. Due to the laws of succession, set S(W) must succeed W, which is the set composed of all the elements W_1, no other sequence facilitates succession. Badiou interprets this in a rather surprising fashion when he says it means that the one-more-step nature of succession can be understood as 'that which hollows out a void between the initial state and the final state. Between the ordinal W and its successor S(W), there is *nothing*. Meaning: nothing natural, no ordinal' (NN 78). Another more provocative way of saying this is: 'a successor ordinal delimits, just "behind" itself, a gap where nothing can be established. In this sense, rather than succeeding, *a successor ordinal begins*: it has no attachment, no continuity, with that which precedes it. The successor ordinal opens up for thought a beginning in being' (NN 78). This is really rather confounding and adds a further counter-intuitive quality to the successor ordinal. What we presumed to be defined as part of a structure without gaps, a pure recursive succession, is nothing of the sort. Each ordinal that we count begins; it has the quality of Nancyean and Arendtian natality. In a very odd way, therefore, the natural transitivity of sets seems to admit no space for something new to occur, yet such sets are founded on the void, as we saw, and now we can say they are structured due to a void. In this gap which says no other element can belong to a successor set except the maximal element of the set in question, one is able to clearly think what a 'real' gap is. A real gap is the gap of pure relational inadmittance: nothing can be found there. It is the perfect example of relational successive limit to such a degree that there can be no *difference* in this instance. If being is differential, it will not be found in this gap which is a non-relational yet entirely related beginning for being every time you take one more step.

Now, in this context, a limit ordinal has no maximal element. It is not the maximum of all the elements of the set w_1 and so it also cannot succeed. Again, this has a surprising and counterintuitive result. Between the 'previous' ordinal in the sequence and this limit ordinal, there would appear to be nothing but void, an infinite gap, but this is not the case at all. Because there is no maximal element here, no w_1 of which this W is the successor and limit, 'between any element of *w* of a limit ordinal L and L itself, there is always an "infinity", in the intuitive sense, of intermediate ordinals' (NN 79). This being the case there is no ordinal closest to it with the result that 'A limit ordinal is always equally "far" from all the ordinals that belong to it ... an infinite distance where intermediaries swarm' (NN 79). We might call this a 'bad' gap for Badiou, because the limit ordinal is unable to hollow out the space behind it of the successor. This transforms the infinite from a value of pure differential and virtual possibility, how it has been taken since Romanticism and certainly within the philosophy of difference thanks to Hegel and Heidegger, into something, like nature, claustrophobic almost. If the space between *w* and its limit ordinal L is infinitely full of ordinals, then the limit ordinal is 'in a relation of *adherence* to that which precedes it; an infinity of ordinals "cements" it in place, stops up every possible gap' (NN 79). Indeed, when one compares the stability of the successor ordinal with the instability of the limit ordinal, it is succession which wins out: 'If the successor ordinal is the ontological and natural schema of radical beginning, the limit ordinal is that of *insensible result*, of transformation without gaps, of infinite continuity' (NN 79).

The section in *Being and Event* dealing with this absolutely essential element is called 'The Second Existential Seal' and tells us little by way of detail as to the nature of the limit ordinal. In effect, all it states is that after the first existential assertion, the void is real, we are now able to make a second existential assertion, the limit ordinal or 'second existential seal' also exists. Hence our long detour into *Number and Numbers* to clarify what I hope the reader can see is a truly essential component of Badiou's work.

Succession

We have already seen how a natural situation presents a mode of dissemination that is homogenous self-presentation. In this sense it is rather meaningless to speak of the dissemination of natural sets. This is also a good way of differentiating successor ordinals from limit ordinals, Badiou notes, through the operator of the union. The union of a set is composed of the elements of the elements of a set. To determine the union of a set, Badiou reveals, you need to 'break open' the elements of the set and collect the products of this breaking: 'all the elements contained in the elements whose counting-for-one' the set assures (NN 79). This fracturing has the apparently paradoxical effect that the union of a set is always smaller than the set, but in reality that is because union moves in the opposite direction to collection. For example, Badiou asks us to take our canonical example of three, a set made up of the triplet: void, singleton of void and pair of void and its singleton. The union or dissemination of the set of the natural number 3 is revealed to be composed first of 0, which gives no elements to the union as it is-not. Then (0), which is the singleton and so can be found in the union. Finally, we have (0, (0)) which is the second element that can be counted by union. Thus, the actual dissemination or union of 3 is (0, (0)) or 2 which is why we say it is the opposite of collecting. When you collect elements together you always end up with +1 element. When you place them in a union then instead you always end up with -1 element. Disseminating an ordinal will never break open the limit if you will of the multiple but only show the elements contained within the multiple: 'the internal homogeneity of an ordinal is such that dissemination, breaking open that which it composes, never produces anything other than a part of itself ... Nature ... can never "escape" its proper constituents through dissemination' (NN 80). The union, the parts of a set when you break it open, is always less than the cardinality of the set as a whole, resulting in the double paradox that when you count a set, close it off, it gets bigger, and when you break up a set, to look at its elements, it gets smaller.

If we now contrast our two ordinals we find that the limit ordinal actually negates dissemination in a manner totally natural. Although it is the successor ordinal that resists being identified with its dissemination, because it must succeed (+1) while dissemination must recede (-1), Badiou defines the limit ordinal such that 'It *is* its own dissemination' (NN80). A successor ordinal 'remains *in excess* of its union' (NN 81) for as we saw, to succeed it must always be one more than the total of its parts. This comes down to the role of the maximal element of the successor ordinal, we have been calling this w_1. W_1 cannot be found in the union of W as it is in fact W + 1. This means that 'the maximal element *w_1 necessarily makes the difference between W and union $\cup W$.* There is at least one element of a successor ordinal that blocks the pure and simple disseminative restoration of its multiple-being' (NN 81).

In that our study concerns indifference naturally of interest is also what determines difference in Badiou's work and here you can see the process of his distancing himself from Derridean and Deleuzian difference. Clearly, the natural situation of total, infinite yet stable dissemination is what he is suggesting is Derrida's flaw, and, for that matter, as we saw Deleuze's. Real difference, then, comes not from the infinite dissemination of differences represented by the limit ordinal's access to an actual infinity of infinities, Cantor's transfinite paradise as Hilbert memorably called it, but from how one moves from one finite number to the next. Badiou confirms that for him the difference between limit and succession is 'of the greatest philosophical importance' (NN 81), a statement to which we cannot help but concur. He notes, echoing his comments in *Being and Event*, that for most it is the limit that has been the fascination, whereas in truth it is succession that is the true mystery. Having dismissed again here Heidegger's sacralization of the limit through his valorization of the poem, repeating several moments in *Being and Event* particularly Meditation Thirteen, Badiou then goes on to detail the complexity of succession in the following terms.

- The ontology of the multiple defines true difficulty not in terms of limits but succession.

- Succession results in a resistance that gives us access to truth through events: 'Every true test for thought originates in the localisable necessity of an additional step, of an unbroachable beginning, which is neither *fused* through the infinite replenishment of that which precedes it, nor identical to its dissemination' (NN 81).
- The success of the limit ordinal in the past is because it has no gaps, but it is the gap of the successor ordinal that is truly profound.

Badiou's conclusion, then, takes these points and combines them into a dictat, I suppose, of true thought: 'There is nothing more to think in the limit than that which precedes it. But in the successor there is a crossing. The audacity of thought is not to repeat "to the limit" that which is already entirely retained within the situation which the limit limits; the audacity of thought consists in crossing a space where nothing is given. We must learn once more how to succeed' (NN 81–2). My simplification of this is that infinity, in seeming to present an exterior to stability, simply reproduces this exteriority to the limit leaving no space for any actual space. In contrast, internally to a set, succession leaves no space between a number and its successor and yet in doing so it opens up a legitimate internal gap. The only place where there is space within the natural set of multiples is locally in the 'one more' that is the essence of the ordinal: to be that which succeeds. Badiou says it differently: 'Basically, what is difficult in the limit is not what it gives us to think, but its *existence*. And what is difficult in succession is not its existence (as soon as the void is guaranteed, it follows ineluctably) but that which begins in thought with this existence' (NN 82). This is clearly in keeping with Badiou's epistemology. What is stable and to be resisted, in that sense difficult, is the stability of the infinity of the natural situation where any difference is indifferently the same. In contrast, what is difficult in terms of succession is not the existence of sequences of numbers but the implications for thought. Resisting existence through the challenge of thought is basically Badiou's credo. It is also where we will find the 'space' for the possibility of the event, not outside or at the limit of a situation, perhaps pushing the current state to an impossible to govern infinity, but within the interstices between each successor element of a state. It is as a state succeeds, in the very nature of the count itself, that a real gap opens up, the impossibility of there being a successor ordinal between two succeeding ordinals. It is in that 'real' gap that Badiou starts to develop his idea of 'real' change, an idea I find difficult to access without the development of the concept of infinity in *Number and Numbers*.

Conclusion on being

The thirteenth meditation concerns Aristotle and the age-old dialectic of the 'already' versus the 'still more' which Badiou reconsiders in terms of what we now know of numbers. Badiou also tackles here certain procedural impossibilities suggested by Aristotle's assertion that any rule for generating the infinite must itself be superseded by the infinite to be a rule 'of' the infinite. Something he considers to be impossible. Finally, he reconsiders issues of number and succession in terms of the big O Other. In that the first issue is one we have certainly covered, that the second is solved by axiomatic recursive reasoning, and the third pertains to Lacan's influence on Badiou, an area very well covered by the literature,[15] I see no need to accompany Badiou any further in this consolidating meditation. Similarly, Meditation Fourteen merely states in a very condensed fashion what we have learnt more expansively from our reading of *Number and Numbers*. There are, Badiou explains, five points from which the infinite is constructed. The first is the mode of passage or succession from one ordinal to the next. In particular, the difference between the point of being or the element as such and the passage from the element to the set of the element or what we have been calling set as collection. The second is the relation between succession as such and limit. The third is the decision: a limit ordinal exists. The fourth is a definition of the infinite by virtue of the assumption, at this stage, of the limit ordinal. The last point is to relegate finitude to being generated by infinity. So, we can say that the infinite is generated in the following manner. First, accept the foundational difference between belonging and inclusion defined by the set as an indifferent collection of elements, not a specific fusion of things. Second, 'count' the sets according to ordinality due to the law of succession by virtue of recursion due to their being well-ordered. Third, define said set of ordinals from the outside as that set of those ordinals using a limit. This results in an interchange between the first existential position, a multiple exists, and the second, a set of multiples exists, by virtue of passage and succession. Now, as regards the limit, propose that if a successor ordinal exists, then a limit ordinal could also exist, that is a set of all natural numbers or a set of all finite multiples. If such a limit exists, can the other rules of set generation, void, succession and limit, apply? If they do apply, then you can demonstrate the first infinite set? This set will follow all the rules we have just proposed. Its point of being, Badiou calls it ω_0 but it is also called just ω and also aleph, does not succeed and is infinite, but every number it generates 'after' that follows the same still more – already structure of passage due to succession. In this way, the first infinite set echoes precisely every aspect of the finite sets which, taken as a whole, *is* the first infinite set. Finally, from this deduct the existence of the finite, solving the problems of defining the infinite as in excess of the finite. This then is a solid place to leave behind the infinite and in so doing also bid farewell to being.

We have come to the end of our consideration of indifferent being qua being. Following Badiou's own summary of *Being and Event* up to this point we can sum up our findings so far.

1. In terms of the multiple, we have shown that all multiples in set theory are content neutral and are therefore defined as fundamentally indifferent.
2. As regards the void, we demonstrated extensively that the void is in-different in terms of its essence. By this we meant cannot even be differentiated. Then in terms of naming the void the 'void' we also showed that the void is indifferent in terms of being without quality and external to the process of multiplicity in the normal sense, as it forms a lower limit ordinal which negates standard modes of differentiation.
3. As regards excess we noted that, in general, the state as such is a negation of indifference and is, in theory, the first moment of difference. That said we also saw that the definition of the state in terms of the relation between presentation and representation is entirely dependent on indifferential rather than actual differential logic. States are different but The State is always the same.
4. The stability of nature by virtue of the infinity of infinities is indifferent as we saw, composed as it is of an infinite multiplicity of multiples. Not only are multiples indifferent but we also traced a powerful relation between actual infinity and indifference due to the indifferent nature of ordinals. This rule counts for both states and nature, which are in effect the same: they are composed of indifferent ordinals whatever their actual named content in terms of rank.[16]
5. Infinity: extensively we showed how infinity is a result of the indifferent nature of sets due to the void, multiples of multiples and their named limit.

Across the five elements of Badiou's ontology which is now effectively over as an object of study for us, multiple, void, state, nature and infinity, it is not simply the case that Badiou mentions indifference, but that the set theoretical and ontological structures he uses are impossible to conceive of without a detailed consideration of indifference. Further, we have argued that a deep understanding of Badiou's ontology is impossible without an equally deep understanding of indifference. Finally, for us, Badiou's ontology is inoperative without a profound dependency on the complex which is indifference. Thus far we can name this indifference as being composed of the following elements:

- Nonrelational in-difference of the void
- Indifferent pure difference as such
- Indifference of all multiples in terms of quality neutrality

This alone is a more detailed menu of indifferential states than any thinker has ever before considered. It is also notable for its valorization of indifference, lacking from Deleuze the main innovator of indifference before Badiou's *Being and Event*, and clearly different in kind to Agambenian indifference, which in parts can actually be said to be used against the dialectical nature of Badiou's ontology. We are constructing a landmark in the history of the philosophy of indifference which we are taking to be contemporary philosophy as a whole. It is clear that, for Badiou, being is indifferent. Now we ask the related question, if this is the case, does it carry that the event, which marks a true state of non-relational difference the like of which I do not think we have seen in philosophy, escape the logic of indifference? Or is it the case that the event, which is dependent on the void of being, is also by definition indifferent? If the latter is the case, can it be that true evental difference can be captured by indifference, a position which seems almost unbearably, perhaps delectably, counterintuitive?

Part Two

Indifferent Events

5

The Event: History and Ultra-One

Having captured being, we advance on the event. 'Event' is a name given by Badiou to designate a real change or an actual difference. He is not the only philosopher to use the term in this way, Lyotard's work on the event is notable for example, as is Derrida's,[1] but he has made it his own. There is a danger which has been realized to some degree that some might take the event to be something that happens that is unexpected and totally original. In this way the event would be akin to the random, the accidental or the revolutionary. Events, in Badiou's work, share these qualities but the random nature of events is not particularly emphasized. The revolutionary or original nature of events is clearly central, but is meaningful only if you take original to mean singular and non-relational, and if you appreciate the systematic, complex and actually heavily proscribed nature of how an event brings about change in a world such as we might call it a revolution. Finally, the relation of the event to revolution makes it, in many people's minds, irredeemably political with a rapidity that cannot be entirely supported. This is particularly the case if one were to take the event as just the most convincing example of radical change philosophy has presented, and so see the event as the crowning glory of the philosophy of difference. Such a strategy could result in an argument easily communicable among our community of saying okay, there were some limitations to thinking of radical change to the dominant structures of our culture as simply the valorization of difference, and these are solved by completing on difference through the singularity of events. If nothing else, my emphasis on indifference ought to show that this cannot be the case. Finally, to think of events as very significant historical moments after which things were never the same again oversimplifies the true nature of the event and raises as many questions as it answers. There is some accuracy in saying that events are historical inconsistencies in the atemporal stability of nature but if they are they're inconsistencies of presentation, not of occurrence. By this I mean that an event is, for early Badiou, a historical singularity, but everything that happens is historically 'unique' in this way, to the point that the singularity of the small 'e' event is lost.

So, an event is not just something that happens but something that happens such that one cannot speak of it in terms that are communicable to the current situation. So ditch the situation? Not quite, for an event is a moment of torsion within a situation, all events occur immanent to situations, so events change a situation from the inside, over time, due to the active participation of quasi-human subjects. All of this being founded on a sense of history in *Being and Event* that Badiou openly rejects in his later work so that saying events are historical moments incommensurable with the state of things is not such a straightforward thing to hold to. Even saying an event is temporal, or that it is something that 'happens' is not so clear-cut. The final word of warning here is that the communicability of the Badiou event in our world is almost entirely based on readings of *Being and Event*, yet it is manifestly clear that the picture drawn of the event there is problematic, so that any future commitment, say, to Badiou's event as a political tool, has to start again with what is meant by the event in *Logics of Worlds*. So that while we will present a basic outline of the event in this volume, the reader will not be fully able to make up their minds on the matter until the second instalment of our study.

It might be apposite at this juncture to remind ourselves that all previous theories of difference wished for difference to be truly singular, and how they failed in their desires. Often this was because the only option in presenting truly singular elements was to define their difference in the abstract, indifferent difference, or to determine the 'different' element as self-enclosed and nonrelationally monadic, in-different. Indifferent difference allows differences to be relational primarily because they are quality neutral, which seems to fly in the face of the idea of a difference. While in-difference makes elements nonrelational, negating the basic condition for determining a difference as actually different by comparing it to the identity of other elements it is different to. To avoid indifferentiating difference, philosophies of difference were forced to consider difference against a background of identity. But to determine identity-differences as regards objects in possession of a unique combination of predicates, one was forced to admit that no difference could ever be truly singular, or the opposite, that any deviation from the norm could result in an event. In the end, we have to conclude, true change remains desirable yet impossible for the continental philosophical tradition after Heidegger. A position ripe for an event, in that the need is communicable but the means currently insufficient.

Badiou's innovation is to concentrate on the stability of being first. Prove being is truly stable, Badiou proffers, and from that gouge out the possibility of a change that is radically new, self-belonging and nonrelational. The only way to do that is to stop thinking of Being as a thing as such, and describe it instead as an operation of consistent stability amongst beings. Set theory allows Badiou to achieve this. In this way you free up the whole concept of the object, thing or element as being an object in possession of a set of qualities. Instead of a new thing, we will speak of a new multiple, and its novelty will not be what it is composed of in terms of its elements, but its operational peculiarity. Events will be multiples which self-belong ontologically and which are nonrelational logically. Or, if you prefer, are multiples whose inconsistency of relation will depend on two differing senses of relationality, ontological rank and logical relation. So no, they are not just unpredictable historical moments after which nothing is the same again.

Finally, embrace indifference. It is, after all, the in-difference of the void coupled with the indifferent difference of multiples in relation to each other in terms of rank, that allows for the ontological statement 'being is-not' to move from impossible to possible for our community. Naturally our one concern here is that if events are first self-belonging and then nonrelational multiples, due to the in-difference of the void and the indifference of all multiples, how will we recuperate an event from being in-different and indifferently different? Will we, in the end, be left with an event defined as radically singular as a mode of real change which either is purely abstract, so not *real* change, or totally isolate, and so is not real *change* in the terms that Badiou takes to be something that can be called change, namely a change in the stability of situations as such?

As we saw in the introduction, Badiou himself has expressed frustration at the way in which *Being and Event* has been interpreted. Concentration on Being has, he feels, been at the expense of the event, while not enough attention has been given to the all-important conjunction 'and'. His point is not that there is being and there are events but how events impact on being. This emphasis produces a three-part project of which we have now covered one element. These three parts are being, event, and then finally the impact of events on stabilities called worlds. In that worlds are not even touched upon in *Being and Event* we have to accept the simple fact that the theory of event presented there is incomplete, perhaps at times incorrect. Badiou tries to convince us of the event using four concepts which, over time, he replaces by the topics and themes of *Logics of Worlds*. These four concepts are evental sites, edges of voids, history versus structure, and the process of naming. In that each is insufficient as we shall see, all the same the four concepts are still able to broach the possibility of thinking a truly singular event, and they allow us to at least show that due to set theory you cannot refuse the possibility of an event just as you cannot present this possibility ontologically due to set theory. Let us say set theory takes us half-way to the event and that that, already, at the time was an immense achievement. But it takes us no further than that; *Logics of Worlds* is needed to complete the task.

Historical singularities (Meditation Sixteen)

Having spent considerable effort explaining what is being-qua-being Badiou, in Meditation Sixteen moves on to consider everything that is not being-qua-being, which is another way of saying the event. This notable shift establishes a fundamental dialectic of *Being and Event*. If there is the natural, there must also be the non-natural, and if there is the non-natural, it is perfectly orthodox to define this as history, the contingent locality. Badiou later accepts this was an unconscionable simplification but for now let us pursue this idea of history as opposed to nature by adding that further, if the natural is the normal state of multiplicities, then all that is non-natural must be abnormal.[2] Abnormality, then, takes us back to the typologies of being delineated in Meditation Eight: normal, excrescent and singular. Of these three, Badiou proposes singular being as the starting point, if you will, of a whole new universe of consideration: everything that is not being-qua-being that is evental.

If you recall, a singular being is one that is presented but not represented. As such it escapes normality, which is the state structure of all representation due to the recount. A singularity is subject to the one-effect, but they cannot be represented as parts as they are made up of elements that are unacceptable to a particular count. In this way, singular beings are undecomposable and thus singular. Badiou explains this by saying such a term is 'not directly verified by the state. Its existence is only verified inasmuch as it is "carried" by parts that exceed it' (BE 99–100). In contrast to a normal situation where all the parts presented are then represented, as regards a singular situation, only some of the parts presented are represented. Badiou gives the example of a family where only some members go out in public. There are innumerable artistic examples of this, think of *Jane Eyre*, *Dr Jekyll and Mr Hyde* and *Psycho*. In each case, a 'member' of the family must always be kept out of the eyes of the state. Such families are, according to Badiou, singular. So, a set is singular when one or more of its presented elements goes uncounted and thus in-exists, in that existence is defined entirely in terms of re-countable multiplicities. This could be taken as being the large numbers of 'citizens' of our states who are totally disempowered by political processes that simply do not include their presence in their legislation.

Badiou says that these abnormal singular states are by definition historical: 'The form-multiple of historicity is what lies entirely within the instability of the singular; it is that upon which the state's metastructure has no hold' (BE 174). To which he adds 'I will term *evental site* an entirely abnormal multiple; that is, a multiple such that none of its elements are presented in the situation' (BE 175).[3] So a multiple made up of some singular terms, we can't call them parts as parts have to be re-countable, is historical. While a multiple made up of entirely singular terms is evental. This early version of the event in terms of singularity is, in retrospect, incomplete. For example, it presents something of a fudge as regards nonrelationality as it opens up the possibility that history is already predisposed to events and that events naturally come out of historical events. We might ask some basic questions here. For example, what is history as such? Badiou defines history as anything that is non-natural, which exceeds the stability of infinite multiplicity. In that said multiplicity is entirely indifferent as regards quality, for most people there is a great deal of abnormality still to contend with. It is evident that something does not quite work as regards the relation of the event to history due to singularity. Much of this is to do with the problematic of what Badiou calls evental sites, or the place where an event can be said to have taken place, a concept he will radically overhaul in *Logics of Worlds*.

An evental site, in *Being and Event*, is an entirely abnormal multiple: 'The site, itself, is presented, but "beneath" it nothing from which it is composed is presented. As such, the site is not a part of the situation. I will also say of such a multiple that it is *on the edge of the void*, or *foundational* ...' (BE 175). His example of this is again the family but one in which all members are 'non-declared' and which always goes together as a single unit. Badiou clarifies his spatial metaphor of 'beneath' here, by saying that beneath the multiple there is nothing 'because none of its terms are themselves counted as one'. It would appear that an evental site is actually a limit ordinal. A limit ordinal is a multiple which does not succeed, meaning there is nothing beneath it. Badiou never says this, which is odd as the reasoning is transparent. An evental site is a place where something is presented that had not been presented before and so it not only contains at least one element that has not been counted before, it is that element. In other words it must be a limit ordinal, either the void or an infinite ordinal.

Badiou goes on: 'A site is therefore the *minimal* effect of structure which can be conceived; it is such that it belongs to the situation, whilst what belongs to it in turn does not. The border effect in which this multiple touches upon the void originates in its consistency (its one-multiple) being composed solely from what, with respect to the situation, in-consists. Within the situation, this multiple is, but *that of which* it is multiple is not' (BE 175). What each of these statements says, in effect, is that an evental site is a limit multiple although it is not clear if all limit ordinals result in events. There is nothing beneath the evental site because it does not succeed. It is the minimal effect of structure in that in belonging to a situation it can be related to other multiples, which is all that structure is. That said, whatever it is composed of up to that point, whatever 'gets' it to the surface of a particular situation, is off-limits because an evental site is minimal or the smallest possible unit there is. The consistency of this multiple is the specific consistency of the minimal unit which so far we have been calling the singleton of the void, although it could also be the singleton of the first number of an actual infinity. Whatever there is in the situation before its becoming, the minimum term for this specific situation is lost to this situation. It in-consists to this situation although it can consist in another situation.

Historical singularities and evental sites: examples

There are problems with history and evental sites yet, all the same, Badiou's vision of situations in terms of singularities is still remarkable. With just a small amount of help from *Logics of Worlds*, for example, we can already present a dramatically enhanced view of the interaction of natural necessity and historical contingency to that available previously. Badiou gives the example of the set {Family}; he also speaks of the set {French Revolution}. This set is an actual and stable infinite that, you will recall, presents at least one singular element which belongs but which is not counted by the overall situation. Now think that a set of multiples called 'family' could also be a multiple of a set called 'social structures'. According to Badiou, there is no limit to these multiples but this is not due to their 'size' per se. Just because 'family' is a part which is included in the set 'social structures' which sounds bigger than 'family' and which could itself be included in a larger structure such as 'human structures' and so on, does not mean that any one of these sets is bigger than the others. Social structures can just as easily be a part of a set: things taught on a British sociology course, for example, and so be much smaller than we initially imagined. The infinity of these sets has nothing to do with the massive size of Hegelian augmentation towards the set of all sets, rather it just means they can never be counted as a whole and complete One. There is no limit to how many sets there can be because the nature of their existence is indifferent. This is not quite the same as saying there are, at any time, an infinite number of sets pertaining to your world.

This is the crucial distinction and the great innovation of Badiou's work. What it shows is that in the field where there is nothing truly transmissible, the field of history or discourse, this lack of transmissibility can be compensated for by the discipline, mathematics, where there is transmissibility. In transmitting a law of consistent inconsistency to the discursive realm, Badiou shows how set theory is not an analogy with how things are in the history of the world. Rather, it expresses formally what we have already realized discursively. What is included in a set is decided by what the set decides to include, not what the things included are, a formalized way of stating our basic axiom of communicability: not what a statement says but that it can be said. This axiom naturally composes a power relation, some agency decides, and so is political, although not in the way that some commentators have presented it. So, to conclude, 'family' can be included in the set 'social structures' as an element which belongs. But that does not mean that any of the things which we included in the set 'family' are available to be counted in the set 'social structures'. It is entirely dependent on the local situation of the first usage of the existential seal 'family' and the second usage of the multiple 'family' now taken as a single multiple of another sealed existential set, 'social structures'. This is what Badiou has to mean by history, the local contingency of every single term as regards its belonging, inclusion, non-belonging or non-inclusion. If we took 'family' as part of 'social structures' and we observed that the nuclear heteronormative family unit was taken to be the base of what we speak of as 'family', we can see that many families are excluded from this designation. What this means is that the multiple 'family' has elements which belong to it which, when the multiple is made a part of a different structure, are not included in this second count. These excluded families or excluded family members become singularized by the process of exclusion, and as such the possible basis for an evental occurrence: a completely new sense of familial and hence social relation.

If only half of the elements of the set 'family' are included in the set 'social structure' it is because 'social structure' could not 'see' the other elements and this is proven by the fact that it didn't count them. Belonging is what is available to count, inclusion is being re-counted. Thus, the hidden family member still belongs to the social structure set, it can be counted as part of 'family', but if it is not counted as one in the new set structure then it is not included. In another twist, the hidden member might be counted and the manifest members not counted, it depends on what the existential seal of the set is: social structures, familial dysfunction and so on. That explains history then for Badiou, after *Logics of Worlds*. History designates the specificity of the relation between sets which themselves are natural, but whose inter-relation is not. Or, history is the localized possibility of the non-natural within a totally natural situation. This means that history is not evental or not yet evental. All it does is indicate that within nature discrepancies occur at the inter-multiple level. In that nature as such is instantaneous and 'eternally' stable, then we call them discrepancies at an inter-multiple level because they can only occur within the contingency of temporality, a point Brian Anthony Smith makes with great clarity.[4] But what about those elements in which nothing of the presented element can be included: the evental sites?

Badiou points out that an evental site is foundational. As we have said it is a limit multiple. There is nothing before it, it does not succeed, we do not know where it came from, it just appeared or was always there but unseen and so on. Sticking with Smith, he explains that, however, the temporal dimension of foundation is the key distinction between natural and evental foundation. He says, reiterating the formalism of the axiom of foundation, 'A situation's foundational element is the one that shares nothing in common with any of its elements'.[5] In a natural situation the answer to this is easy; they are all founded on the empty set, but a historical situation, Smith explains, 'is one with at least one non-empty foundational set'.[6] An evental site then resembles the void in terms of its in-difference: it shares nothing with the other elements of the set because it is non-relational to them in a comparative manner. But whereas the natural set's foundational element shares nothing with the other multiples because it is the subtractive element as regards every count, the evental site is not empty. It contains an element, the void does not. So why is this element non-relational? There must be, from this, at least one other element that is truly in-different which is not the void and the only way this is possible is due to the temporalization of ontology. Something 'comes along' that is non-relational yet non-empty. The nature of this temporality is something we will return to when we consider the generic and forcing but before we leave Smith we should ask the question why the evental site is on the edge of the void. Smith suggests it is because it resembles the void but Hallward's stipulation is better, I think. He explains that the void as such is disseminated nowhere and thus everywhere through a set. In contrast an evental site is a specific location, the first non-empty set due to a void of a 'new' i.e. non-natural situation.[7] The evental multiple, like the normal multiple, must be foundable on a lower limit, a void. The difference being that if you wish you can prove the stability of infinite natural sets through recourse to recursive succession down to the void if you need to, but in the case of the event, it being temporal, you can and indeed must trace its burgeoning stability to the first time, after the void, that this multiple occurred as not-succeeding any other and yet not being the void. This is why it is on the edge of the void. It marks a spatio-temporal break, immediately after the void, just before the first multiple of a new situation. In that the event always occurs in a situation, it also marks an internal edge in that situation, a rupture, that as regards the normal situation, can never exist because, as we saw, there is literally no gap between one ranked multiple and the next.

It would appear that the edge of the void is a way of naming a potential at the 'beginning' of every set and so pertains to the temporal existence of a set or a historical rather than natural situation. Every set starts from nothing, from a void, which separates this infinite set from all the other infinite sets. In that sense then, the evental site is nothing other than the specific point from which a sequential set is extended. Such a multiple can 'enter into consistent combinations; it can, in turn, *belong* to multiples counted-as-one in the situation'. It just cannot 'result from an internal combination of the situation' (BE 175). An evental site then cannot be the void. It is rather the particular relationship of the set to its void as regards how the set began by taking the void set as its first, foundational element {Ø}. This being the case one could say evental sites only occur at the edge of the void, or the first time the void is used to compose a set as minimum limit, not however the first time the set in question is composed. If a singularity in the set 'family' results in an evental site, the set that results is a new set. The point being that the first time the set 'family' is composed is not the only opportunity for an event to occur. We might add that 'first time' is a meaningless designation. As Hallward remarks, the void of a set is disseminated across the whole set, it is not located at the first singleton or lowest level. To which we added that even rank is a construction. Numbers just are, it is humans that insist on ranking them.

The instinct here would be to say that events are new things that happen in history but I find this unsatisfactory. Rather, the history of a situation pertains to its becoming temporalized. This appears to occur due to a conglomeration of several factors. I can count five. First, and most convincing for us as it relates to the order of relation in *Logics of Worlds*, discrepancies occur due to inter-set relations, inclusions and exclusions. When an infinite set interacts with another infinite set in a world, or indeed two worlds intersect or touch, then a discrepancy can occur which is an evental site. Second, natural sets can come to be seen as temporal by the tracing of that set to its foundation, the void, and questioning that foundation perhaps, such that a new multiple on the edge of the void is seen for the first time. Remember that the temporality here is merely one of perspective, seeing a set appear as included in a state as say that which was previously excluded but which always belonged. In that situations are actual infinities, there are an infinite number of elements which potentially belong to the set if we find a way of including them through relation, but which at present do not. This stipulation is not clearly made until the second volume of the Being and Event project. A third temporalization is yes, things happen which means they belong to a situation but they may not be included *yet* or *ever*. The fourth, which we will return to, regards the formal elements of choice and forcing in Cohen's modification of set theory which basically means, in terms of infinite sets of different sizes, there is a potential to force a new multiple. To do this one treats the new multiple in a before and after sense, and this means it has temporality. The final temporalization concerns the relation between natural infinity and the non-denumerable cardinal infinities of the transfinite. As we saw, transfinite numbers can be proven due to the formal procedures that allow us to assert that an actual infinity of infinities exists in a stable fashion (nature), but with one exception, they themselves do not succeed from nature in a well-ordered fashion. Surely this is the real model of the evental site. Evental sites are the temporalization of infinity due to the fact that we can say an infinity of a different size or kind exists, but we cannot say how 'big' it is or what 'kind' it is, and we cannot say 'when' it occurred after the first infinite multiple. Transfinite infinities are infinities with potential, as we cannot name them they could be any kind of situation, and with 'history', they are 'after' the first infinite multiple but not immediately after. If we take this last point and wed it to the first we can say that situations that are temporal occur when there are discrepancies between two infinite worlds due to the fact that they are trans-finite infinities, or worlds whose size and rank we cannot determine yet we know that they do have a size and are well-ordered according to rank. This is, for me, the best guess as to how we could retain the idea of an evental site as a temporalization of a natural situation.

Moving on we can say that this singular, potentially evental, multiple does not succeed, there is nothing before it, which means it is nondecomposable: it contains no elements other than itself. Also, nothing initially at least succeeds from it as its originality comes from it not being available to form combinations with the other elements of a situation. Finally, and this is crucial, an evental site is found within an already existent situation. It is not the same as the foundational or actually infinite interactions with limit ordinals even if it must be itself a limit ordinal. Here is the paradox, an event must be internal to a situation in such way as it is not in combination with any elements in that situation, the only definition in Badiou of being internal.

To sum up, we can now speak of three kinds of situations. The first situation is nearly all situations are infinite and natural: the infinite multiplicities of multiplicities. The second is historical, how infinite multiples of multiples or how a multiple in a multiple comes to be temporalized and thus non-naturalized. Finally, the third situation determines how an entirely new set of elements can only be included if the state, which refuses to include them because it does not see them, is dramatically modified or systematically replaced from the inside by a new set such that this multiple is not only now seen but is seen in such a way as it becomes foundational of a new situation. In *Logics of Worlds* Badiou formalizes these modes of change,[8] but for now we can say than in situation 1 no change as such occurs. In situation 2 a site for change due to singularity is presented. And in situation 3 there is the potential for 'real change' by which we mean the event.

Primal ones and the edge of the void

Badiou names evental sites 'primal ones' and accords them with the following properties. Primal ones are admitted to a count with no previous history of being counted. They are then singularities without history but which are not natural. Structurally, this means these ones cannot be decomposed, they are not made up of parts as this would require at least one level of regression, and these ones do not succeed, so this is disallowed. The result is that they 'block the infinite regression of combinations of multiples' (BE 175) and can act as the foundation for any set of multiples to follow. As they are situated at the 'edge' of the void, before them is the void, after them the singleton of the void, there is nothing below them, literally nothing. In this way, they can found the situation in the same way as the void and the first infinite multiple: 'they interrupt questioning according to combinatory origin' (BE 175).

At this stage we should bring in Hallward's comments on the edge of the void. He clearly differentiates between the basis of all being on the void and the specific nature of the evental site being the edge of the void. 'The evental site is not itself void but that element of the situation that is located "at the edge of its void". The void that sutures a situation to its being is, by the same token, radically aspecific and asituational: it is scattered everywhere throughout that situation. Whatever lies along the edge of the void, however, is always precisely located in the situation. The edge of the void is locatable even if the void is not' (BST 117). He develops this in his later work when he explains that inconsistent multiplicity is the very being of being. As such, as we saw, the inconsistency of being which is-not, is the presentation of presentation as such or that which, in the count, remains uncountable so that the count can count-as-one without recourse to the theological One of total being. Hallward argues that 'whereas inconsistent multiplicity, as the "stuff" counted by any situation, is itself effectively meta-situational, this nothing or *void* is always void *for* a situation' (BST 8). He then explains that the indiscernible elements of any situation, its being, can never be counted as such: 'the void as such remains . . . forever uncountable, forever devoid of any discernible place in the situation. What qualifies as void is scattered indifferently throughout every part of the situation . . .' (BST 8). It is extremely well said. The void as such so to speak is indiscernibly present throughout the situation in an indifferent fashion. It has no location because, as the presentation of presentation as such, it *is* location/situation as such.[9] Or is it better to say that the void as such is the ground of any situation and so in this sense is omnipresent? In any case, the void is indifferent even if it is indifferent to a specific situation. In contrast, the edge of the void or evental site has a place, it is the first in the sequence which does not succeed within a situation.

One final interpretation of the edge of the void can be found in Gillespie's work. He points out that the recount of the state is unavailable to ontology: it cannot think the recount.[10]

Singularity vs. normality

Having determined the nature of the evental site, Badiou is able to explain a few qualities of a singularity. For example, he says that a singular multiple in one situation can be normal in another. What this means is that there is nothing intrinsic to the singular multiple that makes it 'evental', it is entirely to do with its location, hence the term evental *site*. In addition, singularities can always be normalized whereas natural normalities will always be normal wherever they occur, and 'it is impossible to singularize natural normality' (BE 176). The statement which follows: 'history can be naturalized, but nature cannot be historicized' is catchy and provocative but makes excessive use of what Badiou himself calls the banal dialectic of history and nature which dominates *Being and Event*'s view of the evental site but which *Logics of Worlds* moves significantly away from.

Other qualities are also presented here. For example, there is no site in-itself because 'A multiple is a site relative to the situation in which it is presented . . . A multiple is a site solely *in situ*' (BE 176). Again, in contrast to this, a natural situation can be defined intrinsically even if it becomes a sub-situation within a larger situation. Indeed, we saw how each step of a normal multiple is locally global through its combination of transitivity and recurrence. Badiou stresses the local nature of evental sites, arguing that there are 'only site-*points*, inside a situation, in which certain multiples (but not others) are on the edge of the void' (BE 176). A final issue he raises is that while there are evental sites, there is no evental situation. 'We can think the *historicity* of certain multiples, but we cannot think *a* history' (BE 176). Ignoring the use of history here, or perhaps substituting temporalization, the point is still well-made. The event can never become absolutized. Specifically, here he is warning Marxists, I would imagine, against a determinist view of a totalizing history.

Badiou closes this rather controversial consideration of evental sites with the warning that in natural situations there are no evental sites, before adding that the normal situation is only one of a number of states to be found in the 'regime of presentation'. These other states include 'singular, normal and excrescent terms' that bear 'neither a natural multiple nor an evental site. Such is the gigantic reservoir from which our existence is woven, the reservoir of *neutral* situations, in which it is neither a question of life (nature) nor of action (history)' (BE 177).

The main difference between ontology and what is here called historicity is that ontological sets are transitive and historical multiples are not. Put simply, in ontology the inconsistent nature of being means that while the set is founded on the void, this foundation does not begin anywhere. In contrast, a historical situation has an actual beginning, an edge. You can name the specific moment when this set of multiples came into being, this is what is meant by history: the temporal founding of a set. There appears to be a crucial difference here between the founding of the set, and the use of the axiom of foundation to disallow endless regress in the creation of being-qua-being. Being-qua-being is a set whose upper limit is infinite but which has a 'lower' limit which is that it is founded on the void, accepting that this foundation is indifferently distributed across all the multiples in the set at any one time. Badiou provides a proof of this which we can summarize. If we take a non-void multiple α which we say is not an element of itself, then the forming into one of this multiple, its singleton $\{\alpha\}$ shows us that α itself is on the edge of the void for the situation of its singleton or its first representation. As $\{\alpha\}$ does not belong to itself then it only presents one other element which is α and no other elements of α as it presents only one element; all other elements of α are different from $\{\alpha\}$ simply because they are not presented there. This is the basic rule for the composition of all multiples and shows that in relation to every such multiple, as regards the relation of α to its singleton, α is an evental site. Moreover, if we consider the relation of {α}, the situation, to α, the site, we can see that their intersection is void as {α} does not present any element of α only α itself. If then we name what is common amongst the two multiples, what their union is, we find that $\{\alpha\} \cap \alpha$ is nothing, they share nothing in common. This is what Badiou defines as the 'ontological schema of the historical situation' (BE 185) which is 'there belongs to it at least one multiple whose intersection with the initial multiple is void'. This multiple presents nothing in α, is on the edge of the void relative to α, as it presents the void of their union, it formalizes the evental site in α, and it qualifies α as a historical situation. Finally, it can be said that this element, Badiou calls it β here, founds α 'because belonging to α finds its halting point in what β presents' (BE 185) which is nothing at all relative to α.

Self-predication: the matheme of the event (Meditation Seventeen)[11]

There are problems with the evental site as it is presented in Meditation Sixteen reflected in the fact that in the full-scale revision of the event in *Logics*, key issues from *Being and Event* such as history and the evental site being on the edge of the void disappear. That said, if we bracket off these concerns for a moment, the development of the evental site in the next meditation contains several useful explanations as to the nature of the event. Badiou starts by explaining that he is taking a constructivist view of the event because the event is not internal to the multiple so that while it can be localized in presentation, the event as such is not presented nor is it presentable: 'It is – not being – supernumerary' (BE 178). Supernumerary simply meaning taken to be present yet not counted or in some way existent beyond the count. The reason he calls the event constructivist is because usually concepts are presented as constructed and events as what actually happens in the world. This goes back to our warning against seeing events simply as random occurrences within the real world. One of Badiou's significant innovations, therefore, is to say that presentation can be evidentially proven due to the count-as-one, but the event has to be conceptually constructed: 'in the double sense that it can only be *thought* by anticipating its abstract form, and it can only be *revealed* in the retroaction of an interventional practice which is itself entirely thought through' (BE 178).[12] From our perspective this means that while it would be the norm to take the event as something truly different which happens, in fact the event as such must always be approached using the machinery of indifferent thought, namely abstraction and retroaction. The event may be a real difference local to a situation, but by the time you say what that difference is in relation to which situation, the event has passed and so there remains the possibility that, like the void, the event is in-different (non-relational), indifferently different (different only relationally) and finally indifferently abstract (content neutral, available only to abstract reasoning).

The event then is constructed; it is also always localized (BE 178). Badiou explains specifically that this means an event can never concern the totality of a situation, but only a point in a situation, an edge in other words opening out onto the void of not succeeding from anything that is included in this situation. Thus an event always concerns *a* multiple. In explaining what he means by 'concerns' he says: 'It is possible to characterize in a general manner the type of multiple that an event *could* "concern" within an indeterminate situation . . . it is a matter of . . . an evental site (or a foundational site, or a site on the edge of the void)' (BE 178). Thus there is an abstraction of the event, so that while the event pertains to a specific situation, there is an abstraction for all events pertaining to indeterminate situations which are those at the foundational level which we have described. Yet it would be an error to mistake this abstraction with indifference per se: 'there are no natural events, nor are there neutral events' (BE 178). So that while we can access the thought of the event through the indifferent reasoning of set theory, unlike sets, events are not captured by said reasoning and they are not, by definition, indifferent even when spoken of in the abstract. Badiou emphasizes this: 'There are events uniquely in situations which present at least one site. The event is attached, in its very definition, to the place, to the point, in which the historicity of the situation is concentrated' (BE 178–9).

An event then is a construction localized to a specific point in an actual situation. We now consider the relation of the site to the event. Here we must emphasize it would be wrong to attribute any kind of causality between the evental site and the event as such, hard though this is intuitively and also in terms of the way Badiou presents the event. Badiou says very clearly: 'It is not because the site exists in the situation that there is an event. But *for* there to be an event, there must be a local determination of a site' (BE 179). This site is a situation where at least one multiple on the edge of the void is presented. This site is only 'a *condition of being* for the event . . . Strictly speaking, a site is only "eventual" insofar as it is retroactively qualified as such by the occurrence of an event' (BE 179). Such a condition forces us to ask what is the difference between the number 1 {Ø}, the number ω, and an abnormal, singular multiple? All three occur on an edge, yet the number 1 and the first infinite ordinal are both normal, meaning they cannot generate events, while this other multiple is abnormal. Why is this the case if all three are foundational, primal, do not succeed, have a direct non-relation to the void rather than an indifferent distributed relation and so on?

The fourth quality of the event perhaps brings us closer to the answer to these questions, namely that the event is always the result of a self-predicative multiple. This is by far the most important definition of the event as it is the central definition that is carried forward into *Logics of Worlds*.[13] Badiou presents the self-predication of the abnormal or singular multiple in terms of what he calls the matheme of the event. Said matheme is as follows: $e_x = \{x \in X, e_x\}$. Or, he says, take a historical situation X: '*I term the event of the site X a multiple such that it is composed of, on the one hand, elements of the site, and on the other, itself*, or 'the event is a one-multiple made up of, on the one hand, all the multiples which belong to its site, and on the other hand, the event itself' (BE 179).

Why is this so important? Well, for at least three reasons. The first is that when, in the following meditation, he speaks of the relation of the event to the axiom of foundation, he is able to show that said axiom which stops infinite regress as we have seen, while it admits to a void edge, cannot be evental because it is predicated on the simple law of set theory which is that no set can belong to itself. Indeed, the entirety of everything Badiou says about being and set theory is based on the fact that self-predication is impossible. It is the negation of self-belonging that pushes the multiples out from the void, allows them to proliferate in an infinite yet ordered fashion, and means we can also designate using the second existential seal that an infinity of infinite multiples can exist without a totalizing single infinity as ultimate maximum. It is the basic rule that no set can belong to itself that produces the entirety of Badiou's ontology. This means if the abnormal multiple is foundational and primal, it differs from 1 and from ω in this simple way: it is self-predicative. Self-predication makes it possible to differentiate an event multiple, written ε, from the numbers 1 and ω such that we can say the event is a limit ordinal that is not-consistent. This is the only way events, ε, can be possible. Following on from this is the second importance for the self-predicative nature of all events: in that self-predication is literally banned ontologically, it is self-predication that both expels events from ontology and of course shows that events are constructs not beings as such. This ban leads us to a third and final importance which will not be realized in this volume. The primary means by which Badiou is able to show that events are not just possible but actual is the manner in which they operate self-predicatively in worlds such that they can be said to found radically new situations. Events, then, are self-predicative multiples occurring within situations. We can now add the two elements of evental site and events together. Evental sites are non-succeeding sets that, unlike the void, contain an element. Now we can add that this element, unlike the void or the first infinite ordinal, is self-belonging. Events, in sum, are self-belonging and temporalized multiples.

The problem of naming

History, the edge of the void and evental sites are all problematic terms which however reveal important truths about what the event will eventually come to be. The final element to be investigated is perhaps the most problematic, the issue of naming the event through what Badiou calls transcendental indexing. Badiou is speaking of the event of the French Revolution. He examines how in this instance the 'infinite multiple of the sequence of facts situated between 1789 and 1794' indicate the objects of the revolution, in later terminology the logic of its appearing.[14] But that these events do not become a revolutionary event until Saint-Just says that the revolution is frozen which, for Badiou, 'proves that it is itself a *term* of the event that it is' (BE 180). So an event must, according to this model, have a degree of self-awareness: 'that it presents itself as an immanent résumé and one-mark of its own multiple' (BE 180). This self-predication, the set {revolution} includes 'revolution' in the set {revolution}, promotes Badiou into saying that the event is 'supernumerary to the sole numbering of the terms of its site, despite it presenting such a numbering. The event is thus clearly the multiple which both presents its entire site, and, by means of the pure signifier of itself immanent to its own multiple, manages to present the presentation itself, that is, the one of the infinite multiple that it is' (BE 180).

Left like this, revolution is a poor example of the existence of the event as self-predicating. A simple model like saying there are two senses of revolution here, use and mention for example, would explain the double presentation of the term revolution. The issue of course pertains to the problem of naming. With the benefit of hindsight, we can see that Badiou himself comes to reject the 'mysterious naming' of the event as the basis of its self-predication. This is with very good reason.[15] All the same, Badiou carries on with this example raising some of the questions himself as to the consequences of relating an event to its situation. He asks what for him is 'the bedrock of my entire edifice,' the question 'is the event or is it not a *term* of the situation in which it has its site?' (BE 181).

In *Logics*, Badiou dismisses the clear differentiation between event and site, so in that sense we have our answer, but at this stage he simply admits it is impossible for him to answer the question in any simple way. This is due to an intrinsic circularity to the matheme of the event, an inevitable circularity as it pertains to self-predication. Badiou promises that his idea of intervention will solve this circularity but for now let's follow his reasoning as to why it is impossible to say that the event belongs to the situation or that it does not belong.

If the event does belong to a situation, then it is presented. Yet if it is presented it negates the basic premise that the only way a singular multiple can be presented is through the evental site which edges on to the void, and is thus foundationally primate. This is because the event is self-predicating. Because the event belongs to itself, 'it presents, as multiple, at least one multiple which is presented, namely itself. In our hypothesis, the event blocks its *total* singularization by the belonging of its signifier to the multiple it is' (BE 182). In other words, it avoids the indifference of total determination. This being the case, if the event belongs to the situation then it is 'separated from the void by itself' (BE 182). By this it is meant that the event, in being self-predicative, both names the set and belongs to the set so is composed of two elements. Badiou calls this the 'ultra-one': 'Because the sole and unique term of the event which guarantees that it is not – unlike its site – on the edge of the void, is the-one-that-it-is. And it *is* one, because we are supposing that the situation presents it . . . it counts the same thing as one *twice*: once as a presented multiple, and once as a multiple presented in its own presentation' (BE 182).

I suspect the reader may be a tad confused at this juncture and it is such circumlocutory thinking that may have convinced Badiou that the event site was not quite all it should be. Put simply, an evental site edges on to the void but if such a site presents an event, and that is all such a site is, it interposes the event between the site and the void. Between the inexistence of the event as void, not presented, and the presentation of the event, there is the self-presentation of the event which negates the foundational edge of the evental site. In that an event self-presents, if becomes something that can be presented to a world, eventually. This is the difference between the void and the infinite, both of which do not succeed, and the event. The event self-predicates itself into 'existence' such that it specifically cannot belong, ontologically multiples cannot self-predicate, and it is not included, because an event relates only to itself and therefore is not included in the second count. In a very high stakes game, Badiou forces the event to contravene both the laws of ontological presentation and logical, here statist, representation, such that it can be said that there is such a thing as an event.

This sounds like the kind of technical paradox best avoided by simply saying the event does not belong to the situation. But if you said so you would asserting that the event is void because only the void is allowed to be 'present' as site and yet not be presented. Badiou cites Mallarmé, 'nothing would have taken place but the place' but we could also call this the pure presentation as such when there is nothing to present. As we saw, in as much as there is an indifferent pure presentation as such, it is the void and it can only be proposed as such if a multiple is presented. The void is a blank page that can only attain blankness when something is written on it, even if that something is blankness. A blank page with nothing written on it is actually the sign of the void, Ø, but as we saw, this sign is already a multiple, {Ø}, it is not the void itself.

Badiou's analysis of this in relation to revolution is perspicacious: 'if you start posing that the "French Revolution" is merely a pure word, you will have no difficulty in *demonstrating*, given the infinity of presented and non-presented facts, that *nothing* of such sort ever took place' (BE 182). In other words, if language names what already is, then language cannot name the event as event due to Wittgenstein's logic of the impossibility of ostensive language posed in *Philosophical Investigation*.[16] If, instead, you say that revolution is just a signifier within a play of signifiers, then you could never trace those signs back to the original event as said event was itself a pure sign and so on. 'One the one hand, the event would evoke the void, on the other hand, it would interpose itself between the void and itself. It would be both a name of the void, and the ultra-one of the presentative structure' (BE 182–3).

In effect, Badiou is struggling with the fact that within ontology there is a solution to self-predication paradox. The proof that there is no One, only multiples of multiples simply bypasses the self-predication ban by never imposing a transcendental One that is both part and whole of the same set. The problem is because the event cannot present itself ontologically, due to how Badiou designates ontology as natural and normal and the event as abnormal, it cannot participate in this neat solution of side-stepping forever the self-predication interdiction. So it is thrown back into the lion's den of predicative paradox: how can something which never existed be said to exist without relating itself to something which already existed, the ultra-one, or saying it is something totally unique and so is impossible to speak of? Badiou almost shows signs of exhaustion by conceding that one cannot prove the event formally, so instead you need to assert it so as to 'detain and decide . . . By the declaration of the belonging of the event to the situation it bars the void's irruption. But this is only in order to force the situation itself to confess its own void, and to thereby let forth, from inconsistent being and the interrupted count, the incandescent non-being of an existence' (BE 183). There is clearly strategic thinking going on here and you could churlishly say that by defining being mathematically, Badiou makes said being collapse due to the very problem of its own inconsistency. Or Badiou uses set theory only to turn set theory against itself and force it to produce the paradox of the event as proof that there is an event.

Axiom of foundation (Meditation Eighteen)

Badiou now notes that being on the edge of the void, which we are associating with the event as it determines the evental site as a singularity, or an element which does not proceed from another element, is in fact a law of ontology. We already know this but it is worth thinking about it yet again: that which determines the evental site is that it is founded on nothing yet that which determines being qua being is the same. This relates to the somewhat infamous axiom of foundation introduced by Zermelo as one of the last axioms of ZF set theory. This axiom states that every multiple contains at least one site as we know. Badiou says, using our new terminology: 'According to this axiom, within an existing one-multiple there always exists a multiple presented by it such that this multiple is on the edge of the void relative to the initial multiple' (BE 185). It is of course the first of our three laws for generating the infinite universe of multiples: foundation, recursion and the existential seal of naming the set.

So, if we take any set, α, of which β is an element, the union of the two sets is void as no element of β is an element of α. As far as α is concerned, β is nondecomposable. This is rather hard to accept via an intuitive sense of sets. It seems tricky to think otherwise than if β belongs to α then everything that belongs to β must belong to α but this is precisely the case because α takes β as a singleton, so it counts it as one and this is how it belongs to α. In set theory what is contained in a set is what is taken to be contained in a set by a subsequent multiple. It is important to remember that here we are talking of non-void sets which do not belong to themselves or in other words how sets of sets interact, a level we have up to this point not considered. So, while we have already explained the logic of foundation as regards being qua being, nothing can regress below 0, here the axiom of foundation is of a different order altogether as it explains how one set can be founded on another set that is not empty in the first instance. Badiou calls this a 'relation of total disjunction' or alterity (BE 186) and it is important for our study because it is a means of defining difference, relationally, through non-relation. So far, every time we have done so we have called these indifferent differences. If we can show that sets relate to other sets due to a logic of indifference, then we can also show that worlds, which determine the relation of multiples to multiples, are founded on indifference so that they can be actually different and also give rise to real change or absolute difference.

How does disjunction work? First, as regards the axiom of extension, which was crucial along with transitivity in defining sets as indifferent, one set differs from another set if at least one element of a set differs from the other set. This is actually the reverse implication of the axiom which really states if two sets contain the same amount of elements they are the same sets. In contrast to this difference, one set differs from another, the axiom of disjunctive foundation is stronger. It says '*no* element belonging to one belongs to the other'. In terms of their being multiples they have nothing in common: 'they are two absolutely heterogeneous presentations, and this is why this relation – of non-relation – can only be thought under the signifier being (of the void), which indicates that the multiples in question have nothing in common apart from *being* multiples' (BE 186). On this reading, the axiom of foundation, according to Badiou, states what multiples are in terms of their sharing in common their multiple status as a singularity. It simply says you can have two multiples that are related only existentially: they are multiples. Such a formulation is not possible in the world as such, which is determined entirely by relation, and barely in set theory, which determines multiples in relation to other multiples, so it comes into the sphere of an ontological decision. This decision is that there are multiples because there is the void, in this case the void which founds the multiples as multiples and as not other multiples. So, if we have stated that multiples exist, now we need to show that they exist as separate or that there is more than one and less than infinity.

Badiou's description of the axiom of foundation is rather unorthodox. Most take the axiom to say no set can belong to itself, but all Badiou is doing is saying this differently: every non-empty set which does not belong to itself, is radically different from at least one other such set. There doesn't seem to be much which is obviously foundational about this, it is more simply existential, but we will pursue that now. Badiou reads the axiom as saying that every non-void multiple contains some Other, which is the multiple that belongs to it but which shares nothing with it. This new idea of the multiple stipulates that a non-void set is founded inasmuch as a multiple always belongs to it which is Other than it. Being Other than it, such a multiple guarantees the set's immanent foundation, since 'underneath' this foundational multiple, there is nothing which belongs to the initial set. Therefore, being cannot infinitely regress: this halting point establishes a kind of original finitude – situated 'lower down' . . . even though presentation can be infinite . . . it is always marked by finitude *when it comes to its origin* (BE 186–7).

Clearly, this axiom shares a great deal in common with the foundational role of the void for the multiples of mathematics, we spoke specifically about natural numbers; the logic is the same. Where it differs, Badiou notes, is that as an axiom it concerns the ontological nature of sets and multiples themselves so is a meta-theoretical concept pertaining to one key question of the philosophy of mathematics: do sets exist as such? In this way, it conforms to Badiou's concept of the void as the real, because in the end when it comes to any multiple, to say that there is a multiple you need to say that there is a multiple that is not empty which does not belong to itself that is different from another multiple absolutely and which yet shares the status of being a multiple with this radically different, because non-relational, other multiple.

As regards indifference, this has been a particularly tricky part of the history of the term. If we say that multiples are indifferent, in that they are quality neutral, and that sets are indifferent in that they do not gather together elements which are similar but simply name the number of elements that can be collected, then we have a problem as regards stating that there is such a plurality of indifferent multiples. In our schema, it is hard to say there are at least two multiples, let alone that there is an infinity, although in fact if you prove the more than one you end up with an infinity in an instant. For example, if a multiple is different from another multiple indifferently, only in terms of their relationality, how can you be certain that you can establish a relation between these two multiples? Perhaps they are the same multiple. You can establish a relation of non-relation from the void because the void is not a multiple so founding at least one multiple is fairly straightforward. If this logic of absolute alterity works for founding one multiple, taking the void as a singleton, then surely it can be applied to all multiples after that? Each time you speak of a multiple you speak of its non-relationality to the void, and at the same time the void of its relationality to at least one other multiple. It would appear then that if multiples are indifferently different, different only relationally, then there must be absolute alterity in the world, operating like the substrate of the void. In fact, the two are related. When determining the difference of one multiple from another due to succession, you say this is the multiple that it is because there is at least one multiple that belongs to it by sharing the designation multiple, but which is not it because its parts cannot be included transitively. If anyone questions you then, unlike in say other systems of regress, you can trace this back 'down' to the 'first' multiple {Ø} and found that on the void using transitive recursion to rebuild back up whichever multiple you were speaking about. But the axiom of foundation simply says don't bother, all multiples are founded on something they proceed from which is truly Other to it: they share no elements in common but they are in common as element-sharers or multiples.

Implications of foundation

The axiom of foundation raises several very important issues in relation to the dialectic between nature and history that began our considerations here. For example, if, as I have already said, due to the axiom of foundation historical sites exist everywhere (I said they are infinite) then how do we differentiate historical situations which are infinite because founded on a void, and being as such which is infinite because founded on the void? Badiou says this takes us to '*the ontological difference between being and beings*, between the presentation of presentation – the pure multiple – and presentation – the presented multiple' (BE 187). The ontological difference between being and beings is important first because it cuts to the heart of the problem of how ontology and transcendence are linked across the two volumes of *Being and Event*, and because as we saw the presentation of presentation is indifferent, whereas an actually presented multiple is not. As he says, the difference is that in the normal situation, an ordinal, the multiple is founded on the void as such so that for every ordinal the Other names the void and only that. So, a normal situation is one whose historicity is simply the void: it is founded on the void. In contrast, 'a historical situation is reflected by a multiple which possesses in any case *other* founding terms, non-void terms' (BE 188). In *Logics*, this becomes the difference between minimal intensity, and the ontology of their being multiples which can be founded as such. All multiples are founded by the void, whilst all worlds are founded on a void or a minimum point beyond which they cannot go. In this way, they echo the ontological universe in that an infinite amount of elements can belong to a world as long as they fall under the umbrella of the transcendental index, the maximum, but said world cannot get any smaller than nil intensity. Nil intensity is any multiple whatsoever that is not indexed to the transcendental even if it relates to multiples that are. Simply put, the difference between an ontology of pure multiples and a logic of the appearing of multiples in the world is found here. The former is founded on *the* void, so that the latter can be founded on *a* void, a localized point of non-succession. Or, every natural situation is founded on the void, while every other situation, historical here, worlds later, is founded on the void and that fact that it does not contain every other multiple. That this difference depends on indifference is clear. The indifference of the void means that the indifference of the pure multiple as such can give rise to actual differences between multiples that appear in the world due to the presence in each of a radical difference, an Other. Paradoxically, this Other is indifferent because it simply names the ontological fact of there being a multiple of some order. Badiou sums it up neatly enough: 'In the ontological situation, according to the axiom of foundation, to every pure multiple there *always* belongs at least one Other-multiple, or site. However, we will say that a set formalizes a historical situation if at least One multiple belongs to it *which is not the name of the void*' (BE 189).

The last section of the meditation returns to self-predication and finally concedes that the axiom of foundation disallows what Mirimanoff called extraordinary sets. 'But the axiom of foundation forecloses *extraordinary sets from any existence, and ruins the possibility of naming a multiple-being of the event*. Here we have an essential gesture: that by means of which ontology declares that the event is not' (BE 190). To prove that multiples are, you need the axiom of foundation. Said axiom also means that every multiple contains a site of its own singularity as a non-empty set. This would appear to open up normality to an infinite threat of events. If there is an infinity of sites, by definition there is an infinity of events. Yet the same axiom that allows multiples to exist as such, disallows any multiples which are empty or which belong to themselves. Events belong to themselves and are void to any situation as they share nothing in common with any multiple whatsoever, so events are disallowed from the ontological absolutely. 'Ontology does not allow the existence, or the counting as one as sets in its axiomatic, of multiples which belong to themselves. There is no acceptable ontological matrix of the event . . . ontology has nothing to say about the event. Or, to be more precise, ontology demonstrates that the event is not'. But there is also a positive side to this: 'The axiom of foundation de-limits being by the prohibition of the event. It thus brings forth that-which-is-not-being-qua-being as a point of impossibility of the discourse of being-qua-being, and it exhibits its signifying emblem: the multiple such as it presents itself, in the brilliance, in which being is abolished, of the mark-of-one' (BE 190). It was because the void is not presented to a situation that being is-not as inconsistent yet stable multiplicity. And it is because being is multiplicity that what is presented also presents what is not being qua being and which is also not the void. Badiou calls this the event. This is the simple, elegant syllogism of Badiou's philosophy. If there is the void, and there is the multiple, then there is also that which is not the multiple, which is also not the void. This is the event. So we can say that ontology cannot prove the event, it cannot speak of it, but to prove that multiples exist, it must prohibit the event. In this way, if the multiple exists because being is-not, is void, and it only exists if the event is not, self-predicative non-empty sets are forbidden, then it is placed in a kind of hypocritical double-bind. In order to be, multiples depend on what is-not at an ontological level. They must also depend on what cannot be, the event. So for all practical purposes the event has the same status as the void, although the event is not the void. To allow for multiples to exist as such as inconsistent, the axiom of foundation also cannot delimit the presence of the event as similarly something which inconsists so that multiples can. To enter Cantor's paradise, the serpent of the event must first be released!

Although there are problems here with the overall designation of the event we can at least show that events, which cannot be proven, are a part of set theory and thus in this way we can say that they are possible. It is then left to *Logics of Worlds* to show they are also actual. In the meantime, as it stands we already have a complex definition of the event that is pretty workable for many as long as you don't look at it too closely for too long. An event is a limit ordinal by which we mean it does not succeed. It occurs due to certain singular beings which belong to a situation to which they are not included. Badiou calls such situations historical by which we take him to mean singular multiples are possible only due to localized recounts that are temporal. If a member of the 'family' is not included but belongs, then at some point in time they must exist so as to belong but in this instance they do not exist 'enough' to be recounted. As we shall see in *Logics*, attempting to define this quality of available to being but as yet unavailable to beings 'historical' will fail, with the relation defined instead topologically rather than temporally, a major difference in the essence of the definition of the event. From said singularity the possibility of evental sites is proposed. Evental sites are limit ordinals that present to any situation but which are not void limit ordinals, disseminated everywhere, or infinite limit ordinals, located at the maximal point of the second existential seal. Evental sites then are non-successor multiples that appear, literally, in a situation due to their being singular; they can exist but they have not been included in existence. Evental sites are on the edge of the void and in this way they differ from the void itself which is edgeless because disseminated. They are also self-belonging and so they differ from infinite multiples, which are also limit ordinals, because they are the set of themselves and so are not a multiple of multiples but a multiple of themselves. In effect, evental sites allow for a definition of singular multiples as limit ordinals such that they are not the number 1, foundation of all natural sets, and number ω, the infinite stability of all combinatory situations. The fact that evental sites allow for events as retroactively named self-predicative multiples due to historical singularities is modified by saying not all singularities produce events, nor does every evental site produce an event. All that we can say is if there are singular beings, ones which belong but are not included, and nothing of set theory bans this from being the case, then every situation contains at least one singular multiple such that it can have at least one evental site. In short, events constructed from evental sites due to being named self-predicative and because of historical singularities are possible. What they are, and how they operate in actuality, is yet to be confirmed.

Coda: un-relation

It should be obvious now that the event is defined, like the void, but in a different manner, as radically nonrelational, a feature which is only set to be augmented in *Logics of Worlds*' completion of the concept. More than this we have spoken many times of the importance as regards Badiou's ontology of the idea of absolute nonrelationality. We have also shown how indifferent multiples of multiples can be differentiated simply due to quantitative relation, their location within the rank of ordinals, so that their indifferent relational quantity removes the problems attendant on relation due to comparative quality differentiation. Clearly the totality of our comments on indifference depend on the specific use of terms nonrelational and relation in Badiou's work. At the same time, we have hinted that relation as such is actually irrelevant to set theory and to develop this idea, instead of nonrelation, more than once in Part VI of *Being and Event* Badiou speaks of un-relation, a term explained outside the main body of the text in Appendix 2 'A relation, or a function, is solely a pure multiple' where the argument is outlined that in set theory, there are in fact no relations at all.

The thesis is that multiples are multiples, they are not 'objects' with relations, and that while it is commonplace to speak of relations between elements, of sets with relations of order and so on: 'what is concealed behind this assumption of order is that being knows no other figure of presentation than that of the multiple, and that thus the relation, inasmuch as it is, must be as multiple as the multiple in which it operates' (BE 443). There is, in fact, no relation between elements or order of relation as regards sets, there are just multiples of multiples.

Ontology is an entirely un-relational discipline. Badiou concedes that time and again this is forgotten and linguistic terms that suggest relation and order are used. There is, however, a relatively simply technical means for showing that relation can be produced in set theory entirely from the multiple as such. For example, if I say that α has the relation R with β I am actually considering two things. First I speak of the *couple* composed of α and β, and second the *order* in which they occur. For example while $R(\alpha,\beta)$ might be true, $R(\beta,\alpha)$ might not. So it seems like I am both taking elements of a set to be objects and suggesting a relational order between them. Yet in fact, all I am doing is stating two basic laws of set theory. The first is that of the pair or simply 'a multiple composed of two multiples' and the second is the dissymmetry of antecedence that is marked by writing α,β. As we saw, the totality of the natural order of multiples depends on pairing, a multiple and its singleton, and on dissymmetrical precedence, every multiple except the limit is an ordinal defined by the fact that it succeeds.

In fact part of this removal of relationality is dependent on the notation such that α,β is not taken to be the same as β,α: 'the thought of a bond implies the place of the terms bound, and any inscription of this point is acceptable which maintains the order of places; that is, that α and β cannot be substituted for one another, that they are different' (BE 445). This argument has had its detractors but rather than try to unpick their arguments I want to focus simply on Badiou's refutation: 'It is not the form-multiple of the relation which is artificial, it is rather the relation itself inasmuch as one pretends to radically distinguish it from what it binds together' (BE 445). Stepping away from the technicalities of this we can say first that difference can occur in relation to quantity without relation. This is something we have already detailed. The difference between two marks is based on the non-substitutable nature of their placement as regards the ranking of ordinals. They are then different without relation. To this we can add that as regards ontology through quantity, any talk of relationality is artificial, including our own early usage in these pages. There is then no relation as such in ontology, which is why we speak of un-relation rather than non-relation. Non-relation assumes relation as the very basis of differential being and then shows at least one instance where this is not the case. Un-relation however does not assume relation and has no issue with non-relation.

If we accept the above statements then we are able, relatively easily, to demonstrate all forms of 'relation' in terms of multiples and sets. Badiou goes on to show this but we don't need that side of his deliberations. Suffice it to say that it can be shown that relation is nothing other than a multiple, and this includes the function which is simply 'a branch of the genre "relation"' (BE 445). In that we have now explained how it is that relations can be reduced simply to sets of multiples due to pairing and asymmetrical sequential placement in notation, Badiou notes that to work as a mathematician without relations and functions is all but impossible and that these short-hands, which is all they are, are prevalent and necessary if, in fact, artificial and incorrect. As regards this 'technique of abbreviation', however, he goes on to suggest it occurs because being must be forgotten, it does not want to be written. In other words, it is demonstrable of being's inexistence that if one were to work truly on being as such as a mathematician, it would be impossible. So, abbreviations are not just useful, they are revelatory as regards the true nature of ontology. He says 'when one attempts to render transparent the presentation of presentation the difficulties of writing become almost immediately irresolvable' (BE 446). This problem of presentation explains, for Badiou, the prominence of structuralism and the structuralist illusion: 'which reconstitutes the operational autonomy of relation, and distinguishes it from the inertia of the multiple, is the forgetful technical domination through which mathematics realizes the discourse of being-qua-being'. What mathematics forcibly forgets when it speaks of relation is simply: 'there is nothing presented within it save presentation' (BE 446).

All mathematics is reducible to the pure presentation of presentation as such: indifference. Yet relationality means that this fact is forcibly forgotten so that mathematics can function. In that we called non-relationality 'in-difference' we can then now name un-relationality indifferent also in that, as we saw, the pure presentation of presentation as such is the indifferent void at the basis of the presentation of being-qua-being as that which inconsists in the infinite multiplicity of multiples. Badiou says that maths is the staging of beings as objects in relation, instead of being (presentation of presentation). This is why there needs to be a suturing of philosophy to maths. It is only ontologically that pure Being as such as indifferent presentation can be spoken about at all. As regards the language of mathematics, this is simply meaningless (noncommunicable). This being the case, we will continue to speak of the quantitative relation of multiples to other multiples, for now, and will maintain the significance of nonrelationality as well, rather than switch to un-relation, for the same reason that mathematicians speak of relation, as a mode of abbreviation that allows the concepts we are presented to remain communicable amongst us.

6

The Event, Intervention and Fidelity

In that there is an actual infinity of situations or worlds and each of these can entertain an evental site there is effectively an infinity of evental sites. Of this infinity of sites, only a very limited number lead to the appearance of events. The rarity of events is confounding enough, that there is an infinity of evental sites and literally a handful of events, according to Badiou, is troubling to say the least. Having explained the complex relation between an event and its site, such that we can now say that events are possible, we have to come to terms with the precise nature of their occurrence which is profoundly intermittent. Yet the question remains, how can an infinity of opportunities be wasted such that we end up with a paucity of real and actual changes?

To get from an evental site to an actual event, Badiou's rule of passage is straightforward if stringent. An event is a singularity with maximal consequences by which we mean the entire situation that 'results' from the event is maximally determined by not belonging to any current situation and thus exists only in relation to the event as such. Yet Badiou is also clear that an event is not just a novelty or an oddity but is the driver of a set of consequences. Indeed, as we shall come to realize, what is interesting and significant about the event are its consequences. These consequences are formally consistent even if the event which spurred them on is, by definition, inconsistent. For Badiou, an event is a set of consequences resultant in a new situation such that the nature of this situation, its being, is consistent with all other existent situations. There are two possible results to be drawn from this simple formula, both of which suggest that an evental singularity is, paradoxically, indifferent. I say paradoxically because if an event is true change then it is absolute difference and this means, on paper at least, it cannot be indifferent.

The first problem is that which haunts Deleuze, in terms of what Badiou calls the univocity of change.[1] What difference does it make if there is a new situation, if the situation qua situation does not change? Isn't this how the state absorbs revolution and remakes it as reactionary managerialism?[2] The second is that the nature of the evental situation, the consequences felt of an event, reveal something about the state of situations that is permanently destabilizing of our conception of situations per se. If this is the case then the event is indifferent, in that it does not matter what event you are talking about, the consequence is always the same: every state is founded on the void of its own inconsistency. Further, this revolution for the state is cyclically ineffective as clearly it forgets about its own inconsistency as the consequences of the evental situation become slowly recuperated by the representations of the state. As Badiou is at pains to negate the Marxist view of a final destination for history, and now in these meditations also an apocalyptic founding moment as well, then it must be the case that all militant subjects for truth become grand old women and men of the establishment. Whichever of these two eventualities is closer to the truth, the event is indifferent and discursive not unique and ontological. These observations do not negate the possibility of the event but they do force us to reconsider what the nature of an event is, due to its rarity placed alongside the infinity of sites and situations it emerges from.

We can better consider the event's existence and its rarity if we linger on Hallward's brief summary of how an event comes to have consequences. He says that a truth, truth is the name of the consequences of the event, is not 'the void made present' (BST 122) but is 'constructed bit by bit, *from* the void'. Specifically, it is constructed from the means by which evental sites found the events which precede them, through the intervention of a subject. Putting aside, for now, the question as to what this subject is, relative to other ideas of subjectivity, we can say that the subject's job is to name the event so as to make it represented within a situation. Then, they must intervene on that name to 'make it stick' (BST 124). Next, a subject decides which elements of a situation affirm this name and which do not. And finally, they work towards the founding of a lasting fidelity to this named evental situation. What we can draw from Hallward's analysis is that for an event to stick and produce consequences, in other words for it to exist as a situation and not just a site, we need to have a subject who is committed first to intervention, naming the event, and then fidelity, constructing consequences of this name. Intervention and fidelity explain both how there are events at all, and also why there are so few.[3] In simple terms, intervention requires courage and fidelity requires stamina, which appears to be a rare combination for any epoch.

And so it is that the fifth section of *Being and Event* busies itself around the twin topics of intervention and fidelity. In that these are based on a subject due to evental sites, some of these comments will be modified when we turn to *Logics*, but the essence of the role of the subject is retained across the two studies.

As fidelity deals with consequences, it is inevitable that we will not be able to tackle it without reference to the logic of appearing as this determines Badiou's theory of consequences presented in *Logics of Worlds*. This means we will have to extend our understanding of consequences beyond Hallward's introduction, but for now, however, we can consider intervention solely from the confines of the earlier work with the proviso that if the subject names the event, then please keep in mind that the later work questions the assumed power of naming and so that agency will have to be replaced.

Keeping company with Hallward a while longer, he defines intervention as having two parts: courage in naming the event and 'determination to make this implication apply' (BST 125).[4] There is, as I have said, a clear element of discourse implied here. The process of nomination is central to Foucault's statements and Agamben's paradigms.[5] In addition, the ability to make an implication apply concerns force, which pertains to power. There is a sense, therefore, in which Badiou is suggesting, in terms of the event, nothing more complicated than the ability of truth to impinge upon communicability. The question is, how truthful are his claims, by which I mean to what degree can he determine a nomination that is distributed through time that is different from any other discursive statement? For example, any name that enters discourse is a new name. If it were not, it would already be included in the conditions of communicability of said discourse. So, in Badiou's terms, the new name must be new to communicability as such. Not just inconsistent to the discourse of European power, but inconsistent to the very rules of discursive communicability said power is based on.

If we use the example of Kant's communicability as regards the aesthetic in the second moment of the *Third Critique*, in traditional communicability when asked by a subject: Do you find this beautiful? the other subject is free to answer 'yes' or 'no'. The content of their response does not matter as in either case they confirm that they understand the terms of the question and thus that they share a common subjectivity with their interlocutor. In contrast, in Badiou's system when the question is asked after the use of a nomination, what is questioned is not a new object in relation to an existent set of names, but a new name in relation to an existent set of situations. This being the case, contra Kant and Habermas, subjectivity occurs out of a lack of communicability.[6] It is not the product of a communicable action. Subject A asks subject B do you find this world *X*? Subject B does not know what *X* is, so they can neither answer 'yes' nor 'no'. Thus, subject A is truly a subject as she is concerned with a truth, and subject B is not, unless they can become concerned with that truth, which up to this point they were unaware of. How they become concerned with a truth requires not that A explain to B what *X* is, but the demand of subject A on subject B to leave the consistency of their discursive world and join A in an as-yet inconsistent process of step-by-step world-making.[7]

Accepting that a name is a new name, for it to have evental consequences it cannot be legitimized by the power that already exists. So it's not enough to have a new name, you also need to establish a new power for that name that does not emulate in any way current forms of power, and indeed destroys them. This is the only way that Badiou's evental interventions can occur.[8] This second point concerns fidelity through consequence so we will not pursue it further yet. Instead, we need to analyse the specific nature of intervention in answer to Hallward's convincing assertion: 'Intervention is purely a matter of yes or no, it did happen or it did not happen, and this yes or no applies only to the existence of the event, rather than to its alleged (and always debateable) "meaning" or manner' (BST 125). This is perfectly in accord with Badiou's own comments on intervention which come down to one simple maxim: you must decide or choose as regards the existence of the event because, in terms of being, the event in-exists. Yet it also contravenes our law of communicability, which is being able to say 'yes' or 'no' to keep the name of the event within the field of address of the discursive world in question, meaning that the name of the event must be something to which you can say neither 'yes' nor 'no'. In other words, the nominative act of the event is singularly indifferent. This is not necessarily disagreeing with Hallward, as much as emphasizing a different approach to the question based on different ends: Hallward wants to establish a subject to truth, we want to determine the relation of indifferent being to singular, non-neutral, situation-specific events.

The wager: yes or no (Meditation Twenty)

Meditation Twenty starts with a recursive glance back to the question as to whether an event does or does not belong to a situation. As we saw, neither position was tenable in the mathematics of set theory, and so here Badiou asserts that the undecidability of this question of belonging means that you must decide or make a wager. The choice of the term 'wager' refers back to the previous meditation on Mallarmé and the throw of the dice, and presages the forthcoming meditation on Pascal.[9] We will leave those sections to the perspicacity of others to be glanced over at the reader's leisure.[10] Suffice it to say that when an undecidable situation occurs such that it gives us 'no base for deciding whether the event belonged to it' (BE 201), then we must decide for ourselves.[11] This imperative is an essential part of Badiou's ontology and is why we call his philosophy decisionist.[12] That said, we do not decide if an event belonged or not, rather we decide if there was an event or not relative to the impossibility of saying either way in relation to the situation. This stipulation suggests that the inability to say 'yes' or 'no' is an indifferential suspension between the existence or not of something in relation to being as such in its normal situation. It does not follow that undecidability is the same as indifference, for example Derrida's wide-spread use of undecidability never becomes indifferential,[13] but in this instance undecidability pertains to whether an event even occurs, and so it suspends the yes/no structure. For example, if you say 'yes' to the event, this is not the same as saying 'yes it belongs', because as we saw it cannot. Just as if you say 'no there was no event' this is not the same as saying an event does not belong because such a non-belonging event does not take place. Instead, when you say 'yes' to an event you simultaneously say 'no' to the terms of how multiples can be said to exist, and this is the crucial difference.[14]

To further clarify this matter, Badiou says that if you do say 'yes' to the event your wager, which is illegal, must never become legitimate 'inasmuch as any legitimacy refers back to the structure of the situation. No doubt, the consequences of the decision will become known, but it will not be possible to return back prior to the event in order to tie those consequences to some founded origin' (BE 201). In addition, the process of decision on the event 'requires a degree of preliminary separation from the situation'. The situation can't provide the means 'for setting out such a procedure in its entirety. If it could do so, this would mean that the event was not undecidable therein' (BE 201). There is no 'regulated and necessary procedure' which works for the decision as to the 'eventness of a multiple' (BE 201), which confirms our own assertion that as regards the discourse of the naming of the event, the naming must question the laws of communicability as such, otherwise said name is always interpolated into discourse even if it is included in a negative fashion. It is not a matter of saying 'yes' or 'no' to the event, but of saying 'yes' to the event because you cannot say 'yes' or 'no' to the event as belonging. If you say 'yes' to the event, you say 'no' to the ontological laws of being, a decision which describes the specific process that Badiou calls intervention.

Intervention

Badiou defines an intervention as 'any procedure by which a multiple is recognized as an event' (BE 202). For such an event to be recognized he says two things are implied. First, that the form of the multiple is said to be evental, by which he means is formed out of represented elements of its site *and* itself. And second, that as regards this multiple, you decide that a term of the situation belongs to said situation. Another way of putting this is that one identifies an undecidability due to the self-predicative nature of the multiple, and then decides on the multiple belonging to a situation. Yet there is a problem here in that the second point appears to cancel out the first. If the event is undecidable in form and you intervene on it through deciding, you, by definition, negate its form and disallow the event. Citing similar issues presented in his reading of Mallarmé in the previous meditation, Badiou calls this the event's 'auto-annulment of its own meaning' (BE 202). The deadening auto-annulment of an event leads to a key paradox surrounding the action of intervention as decision: 'what it is applied to – an aleatory exception – finds itself, by the very same gesture which designates it, reduced to the common lot and submitted to the effect of structure' (BE 202).

Badiou goes deeper into this 'paradox of intervention' (BE 203), when he explains that the two elements that seem to cancel each other out are in fact 'inseparable'. This pertains to an element of nomination that Badiou does not address directly but which seems to be part of his overall conception of nomination: deixis.[15] Evental deixis is entirely exhausted by its referentiality: this event took place. Yet by the same gesture it contains no specific content and so is purely material without any specific reference. It is, in other words, an indifferent form of nomination wherein a content-neutral word is employed in a context where the being of the thing named is such that it can easily be assumed. If we say of an event, *X* is an event, we are in effect using the opposite of deixis. Instead of a general signifier, 'this', which is meaningful due to a consistent context, we have a singular signifier, [*e*], which is truthful due to its lack of consistent context. Yet at the same time there is a real danger that the nomination of the event could succumb to the logical impasse of deixis simply by entering it from the opposite direction. Badiou explains that when we say that the event of site *X* belongs to it, e_x belongs to e_x, by recognizing *X* as a multiple the formula 'supposes that it has *already* been named' (BE 203). Naming the event does not determine its reality but it does suppose it to be susceptible to a decision as regards whether or not it belongs.[16] 'The essence of the intervention consists – within the field opened up by an interpretative hypothesis, whose *presented* object is the site (a multiple on the edge of the void), and which concerns the "there is" of an event – in naming this "there is" and in unfolding the consequences of this nomination . . .' (BE 203).

It cannot be enough to say there is an event. The indicative nature of deixis requires a specific context wherein you are legitimate in using the generalized 'there is' or presentation, because the context of this indication is already secure. What Wittgenstein calls the problem of ostensive thinking is also in another, literary, register called anaphoric deixis. Yet there is also a tendency towards cataphoric deixis in poetics and art in general, the opening of *Paradise Lost* being a famous example. In cataphoric deixis, a general designate is used first, and then the name of the person or thing referred to is added in later. This protensive mode functions because the contextual nature of appearing is presupposed, even if what will appear is not yet known.[17] Certainly, as regards intervention the name is an example of cataphoric deixis. The event occurs and then is named as having occurred. Yet in this instance it is not a delay indulged by consistent context, it takes Milton many lines to get around to the subject of his epic, but the possibility of deciding on a complete lack of context. Also, as we said, in actual fact the naming of the event is anti-deictic as it uses a unique name with no referential context. But if evental nomination is not deictic, nor can we call it pure ostensive reference because it does not point to the event as such. If you name an event you do not name the thing itself, which in any case is not a thing, but you actually show your willingness to attest to its having occurred. In this way Badiou's anti-deictic, cataphoric nomination is a speech act. The subject does not name a unique thing with a unique name, rather they name themselves into being a subject due to their decision as regards the event because they have chosen to name it. Specifically then, the subject *is* the result of a neo-Kantian communicability and quasi- or anti-Habermasian communicative action after all.[18] When the subject names the event, they do not tie a predicate to an object or concept, instead they make themselves into a subject through the retroactive naming of an already passed event.[19]

Perhaps it is helpful to look at Badiou's own consideration of the meaning of nomination. '[W]hat resources connected to the situation can we count on to pin this paradoxical multiple that is the event to the signifier; thereby granting ourselves the previously inexpressible possibility of its belonging to the situation?' (BE 203) he asks before moving through several possibilities which he rapidly refuses. Nothing from the situation that is presented can do the job as it would efface the unpresentable nature of the event. The site can't name the event for the site *is* a term of the situation and so on. These are just reiterations of the various problems we have already addressed that put off the moment of the final, founding maxim of intervention: 'The initial operation of an intervention is to *make a name out of an unpresented element of the site to qualify the event whose site is the site*' (BE 204). This nominative act determines a specific kind of naming which he describes in the following fashion. From this point on the *X* which is the name of the event or better 'indexes the event e_x' will not be taken as *X* (that which names the site) but an $x \in X$ that *X* counts as one in the situation without that *x* being presented in the situation. He says: 'The name of the event is drawn from the void at the edge of which stands the intra-situational presentation of its site' (BE 204). So, we can say that *X* does not name the event but names the ability to count as one said *X* without it being presented. This is clearly a form of retroactive cataphora and, as it concerns indexing, there is a deictic element therein, but it is not a cataphora which can be made meaningful by referring back and saying this *X* is the name of an event. Nor can one say that this *X* points to an externality without remainder as the *X* in-exists. Rather, the name of the event is actually the oath of a subject to be true to an event, or a performative form of speech act that enables the ability to speak without consistent context in a manner that will become communicable precisely because it negates the basic laws of communicability.

Seven consequences of the event

Badiou details the seven consequences which are the purpose of the act of intervening so as to 'name' the event. All the same, irrespective of the detail, in *Being and Event,* the term consequence is somewhat ill-defined formally. This shortfall is rectified in *Logics of Worlds* where Badiou explains precisely what he takes to be a consequence as regards the site. If we take two transcendental indexes, for now take those to be the names of two sets, then we say that the first 'depends' on the second, let us call them q and p, if the degree of intensity of p to the same index as q remains lower than or equal to q. This is based on the simple logical operators of appearing greater than, lesser than or equal to. If p is less than q, then q depends on p to the degree that p is lesser than q. Dependence describes the relation between transcendental degrees by explaining the elements they share in common, and which of the two transcendental degrees is the greater. If q is greater than p, then it depends on p for all the elements up to the size of q. It does not make these elements part of its world all over again, rather it includes them as already appearing. If we consider this structure, then a consequence describes the relation of objects in p (the lesser) and q (the larger). So, if an object, a, appears in p and another, b, appears in q, if we consider the relation of dependence of p and q that will also tell us to what degree object b is the consequence of object a. All this really gives us is a logical means of determining how relations between groupings carry over their intensity of relation as regards the objects they group. As Badiou concludes, the point here is that 'consequence is a (strong or weak) relation between existences, and that therefore the degree according to which a thing is a consequence of another is never independent of the intensity of existence of these things in the world in question' (LW 371). Consequence determines a causality based on the relationality of two elements under the auspices of the third they share in common. As regards the consequence of an event, this impacts on the world and then on being as Badiou explains with a clarity and elegance missing from his comments on the event in *Being and Event*:

> When the world is violently enchanted by the absolute consequences of a paradox of being, all of appearing, threatened by the local destruction of a customary evaluation, must reconstitute a different distribution of what exists and what does not. Under the pressure that being exerts on its own appearing, the world may be accorded the chance – mixing existence and destruction – of an other world. It is this other world that the subject, once grafted onto the trace of what has happened, is eternally the prince (LW 380).

Having detailed what Badiou comes to take as a logical consequence, the final word on evental consequence, we can now return to the septipartite definition of evental consequences Badiou presents in his earlier work. Many aspects of his description of evental consequence will be superseded in *Logics of Worlds*, yet all the same the long debate on consequence in *Being and Event* reveals an immense amount as to the relation of the event, as consequence, and indifference.

a. Concerning the matheme of the event, we can say that the proposition $x \in X$, or self-predication, has two functions. It is both the unpresented element of the presented one of the site at the edge of the void, and it links the event to an arbitrary signifier. All intervention is composed according to this 'double function' meaning, as we saw, that its maxim is '*not tied to the one, but to the two*' (BE 205).

b. Now we turn to the name itself and the problem of its specificity. In particular, how does the name of the void differentiate itself from the void when, as Badiou says, 'the law of the void is in-difference' (BE 205), or the inability to be differentiated? More than this, the name of the event is 'in itself, anonymous. The event has the nameless as its name' (BE 205). All the possibilities of the naming of the void being otherwise than this are considered with the usual consequences, namely the manner in which the count-as-one would negate the unpresented nature of the event. Badiou is left with an ascetic fact that all you can say of the name of the event is that it belongs to the site: 'It is *an* indistinguishable of the site, projected by the intervention into the two of the evental designation' (BE 205). While this anonymous name is on the way to becoming differentiated, its first function, as we saw, is a kind of ostensive indexing or negative deixis. As such, in its indistinguishability it remains in-different right up to the point of its consequences.

c. We now consider the name as regards the state. Said nomination is illegal according to the state as it does not conform to the laws of representation, something that belongs which is then included. As we know, the singular belongs in such a way as it cannot be represented. The choice of intervention on the part of the subject is described by Badiou as a non-choice by the state. One can concede that the term of the site that names the event is a form of representation, but we have already seen that it is closer to an indexing, denotation or ostension than any normal form of representative naming. Badiou says that as regards the state, this mode of representation cannot be recognized therein: 'Because no law of the situation thus authorizes the determination of an anonymous term for each part, a purely indeterminate term . . . a representative lacking any other quality than that of belonging to this multiple, to the void itself . . .' (BE 205–6). The state, in short, has no means of recognizing indifference of this order, which names an element purely in terms of its relationality, here its non-relationality, irrespective of its qualities. We can say then that the naming of the event is a means of extending the radical nature of indifferent being into the world of differences. Not only does the self-predication of the event allow being to interpose on the world through the effects of its paradox, it also permits the indifference which dominates the ontological to make its presence known in the communicable worlds of differentiation.

d. Now Badiou puts these elements together, specifically the anonymous nomination of the event, its pure indifferently deictic 'there is', and its status as the ultra-one obeying the law of the Two not the one. Badiou notes that the naming of the event is not possible in the law of representation. This means that intervention only works as a means of 'endangering the one'; it is felt in terms of its negative effects within representation. The term of the event is not a term in the same way as other terms of the state are terms. Only as regards the event does the term nominate the void and thus is never the name of another term, but an actual event. This intervention established the specificity of the event, its one-ness, as a non-one or a one *in absentia*. Taking the name of the event as [e_x] one can say that the event is this event, but 'inasmuch as its name is a representative without representation, the event remains anonymous and uncertain' (BE 206). This is one of the apparent paradoxes of indifferent designation.

We have already seen that specificity of differentiation can easily occur indifferently due to the logic of abstract relationality. That said, the specificity of the two terms differentiated due to relation were underpinned by their place within the consistent structure of nature, that is in terms of their relation to the void, the term which it succeeds and the upper seal of an actual infinity. Here there is, first of all, no relationality. Now the void is in-different due to non-relationality and the evental term is indifferent because of the void, but remember the event is a non-void term. The evental term's indifference is not fixed in place by the void, it does not participate in the relational yet indifferent world of natural numbers say, and it has no upper bound in that it is both minimum and maximal at the same time. The evental name then comes close to one of the longstanding problems of language: the pure denotative function. What this reveals is the pure presentation as such of presentation or the linguistic function per se. You can name but you do not say what you name.

There are problems attendant on this process but we have already shown that Badiou moves away from nomination so we will leave it there. Instead, let's think about the last comment of this section. Taking the essence of the ultra-one of the event as Two, its placement between the void and *X*, Badiou is able to say that 'an event is an *interval* rather than a term: it establishes itself, in the interventional retroaction, between the empty anonymity bordered on by the site, and the addition of the name . . . The event is ultra-one . . . It is an originary Two, an interval of suspense, the divided effect of a decision' (BE 206–7).[20] The event intervenes, if you will, between the nameless and the name and this is its function. The event is not something new or unique, instead it is an intermittent moment[21] when the stability of the ontological and of the world of appearing are placed in a kind of relation by which the paradox of being is displaced into the stability of its appearing in such a way as the world is creatively 'destroyed'. Badiou says in this context several redolent things. For example, he proposes the possibility of an originary Two. Here we can differentiate any world as such, from that as a consequence of an event. Any world as such is founded on the counting as one of the void resulting in the being of indifferent multiples of multiples in an absolute fixed order. A world which is not founded on the void as such, but on the interval between the void and the one, breaks the ontological law of all worlds. By rights these worlds should never be of the same order as all others. There should be clear differences between the relations of appearance of a two-founded world and those of a one-founded world. Alternatively, perhaps the point is that all one-founded worlds are two-founded in reality, and the event simply shows that, negating the stability of the very idea of a world. Note that in each case the event cannot touch being so we are always talking about the effects of thinking about being on the worlds we have built upon its stable laws.

The next point is also very important. The interval is not liminal, it is not dynamic, it does not establish an economy, an oscillation or any form of actual relationality. It suspends, rather, the assumed ease of relation between void and one and for me this is the best way to think the event.[22] The event is what suspends, due to its anonymous indifference, the easy relation between the void as such and its presentation as a singleton. The event demands that the one, the first singleton, succeeds not from the void, but from the event. This must have disastrous consequences which Badiou on the whole does not detail. Simply put, the whole basis of a founded, transitive and recurrent ontological stability needs to have no thing between the first one and the void, indeed between every one and the void. The last statement simply makes clear that Badiou's reliance on the Two of the event means his will always be a decisionist philosophy. We will return to this when we look at the axiom of choice.

e. Badiou now says that intervention as such is also undecidable and is recognized in the site by its consequences. The actual occurrence of an event will always be doubtful, an act of faith on behalf of those who intervene: 'What there will be are consequences of a particular multiple, and they will be counted as one in the situation, and it will appear as though they were not predictable therein' (BE 207). So that while there always will have been some chance in the situation, the one who intervenes can never claim that this chance is the result of their intervention, even if the event is named retroactively due to their intervention. 'Intervention generates a discipline: it does not deliver any originality. There is no hero of the event' (BE 207). Again, the event is nothing new. The nature of the event, its indifferent anonymity, means the event as such is always the same. What is new are the consequences but the fact that there are consequences is not new either, intervention as such is not novel. This is why intervention is a discipline. It is a systematic mode of making sure there are consequences to the event that are presentable to a world to such a degree that they destroy a world and make a world. All these effects are new, but the origins of these effects are indifferent, abstract and fixed. This is obvious and Badiou is brave to say it. In that the event is a consequence in a world of the paradox of being as regards self-belonging, because being never changes, the fact that there are events can never change either. There is a positive side to this: the absolute consistency of nature guarantees the absolute presence of inconsistency. And a negative side: the event is ultimately a servant to nature and can never disrupt nature except by negating itself permanently.

f. Now we come to the whole point of the event. Badiou says that the state, presented with the consequences of the intervention on the evental site through the act of naming, needs to recuperate this name or face destruction. I can include this multiple into the state: 'at the price of pointing out the very void whose foreclosure is its function' (BE 207). The state, then, has to come to terms with the parts of the evental site, for the state is simply the mode of forming parts out of terms which belong (this is the set theory definition of inclusion – sub-sets). At the same time, it needs to be able to tackle the self-belonging of the event meaning its only recognisable part is itself! The state needs to deal with the site, which we will name X, and the forming-into-one of the name of the event or $\{e_x\}$. Accordingly the state names the event thus: $\{X,\{e_x\}\}$, the site plus the forming into one of the event through naming. Obviously, the name of the event as regards the state is a Two or X plus e_x = the event. 'The problem is that between these two terms *there is no relation*' as the name of the event, as we saw, is illegal in the eyes of the state so the name of the site cannot be represented. The state can register the effects of some novelty, some glitch in the system, but that is as far as it gets: 'From the standpoint of the state, the name has no discernible relation to the site' (BE 208). The state often represents this in terms of enigmas, paradoxes, the effects of strangers and the like and indeed these, we presume, are good places to look for evental sites.

g. Badiou ends with a consideration of the problem of evental circularity. 'It seems that the event, as interventional placement-in-circulation of its name, can only be authorized on the basis of that other event, equally void for structure, which is the intervention itself' (BE 209). Or, there is no intervention until there is an event, but there can be no presentable event unless there is an intervention. 'It is certain that the event alone ... founds the possibility of intervention. It is just as certain that if no intervention puts it into circulation ... then, lacking any being ... the event does not exist' (BE 209). The only way out of this hermeneutic circle, for Badiou, is the radical and perhaps troubling suggestion that '*the possibility of the intervention must be assigned to the consequences of another event* . . . An intervention is what presents an event for the occurrence of another. It is an evental between-two' (BE 209). He appears to be saying that one event follows another, but does this mean, as Hallward suggests, that there is then some degree of relationality as regards the event?[23] If there could be no French Revolution as intervention without there having already been the felt-consequences of another event, say the American Revolution, then events form a causal, relational chain do they not? This is not what Badiou says, although he leaves this issue woefully underdeveloped here. So that while he says that the event of an event means that events form the basis of a theory of time: 'for there to be an event, one must be able to situate oneself within the consequences of another', this temporality, however, cannot be traced back to the very first event or projected forward to the very last (à la Marxism). He leaves the issue there. Perhaps we can fill in some gaps by saying that as the event 'depends' on ontology for its coming to being, for its existence basically, then it is party to the same laws as multiples as such. As we saw although we present the stability of multiples as a sequence, in fact the void determines every multiple indifferently and it is only the event that has a named, specific place for its relation to the void. This being the case as no structure of multiples begins anywhere, so no first event has to, or indeed can, begin all others.

As regards the causal link, at best it is only quasi-causal. All Badiou is really saying is that interventions are possible because of the consequences of other interventions that occurred before showing that there can be interventions. Yet his invocation of time, and his use of innovations in mathematics as an example of this, makes it very hard for the reader to come to any other conclusion than Hallward's namely that as regards the relation between one event and another, there is a relation. I find this a highly troubling eventuality and suggest that at this point Badiou does not solve the problem of circularity by saying one event depends on another. In my opinion it is enough to say, atemporally, that in as much as there is structure, there are always events and so there can always be interventions meaning there can always be other events. But to say this is a temporal function seems completely incorrect. It is simply the effect of structure as such. Badiou here seems trapped in his insistence that events are historical not structural, a problem we saw he removes in the later concept of sites that can appear without the problem of naming. So we can say at this point if the event is the result of an intervening nomination, at some point the problem of historical relational succession will occur, so the means of avoiding that are the refusal of the dialectic of history and structure and a movement away from naming the event such as we find in *Logics of Worlds*.

For me this eventuality negates the closing comments of this section but it would be neglectful to ignore them even if our position is that they are wrong. Badiou says: 'the event itself only exists insofar as it is *submitted*, by an intervention whose possibility requires occurrence – and thus non-commencement – to the ruled structure of the situation; as such, any novelty is relative, being legible solely after the fact as a hazard of an order. What the doctrine of the event teaches us is rather that the entire effort lies in following the event's consequences, not in glorifying its occurrence . . . Being does not commence' (BE 210–11). The last maxim gives us the clue as to the legitimate reason of the relation of one event to another. And the insistence that an event is not something new but a set of new consequences is laudable. Yet saying novelty is relative runs the risk of making the event little more than a version of Derrida's idea of invention of the impurity of singularity.[24] Badiou is clearly trying to come to terms with the retroactive nature of the event, yet one wonders why he needs history for this. As we saw, the axiomatic method is by definition retroactive and while this is in a way 'historical', axioms become accepted over time due to the interventions of subjects, the reason they become accepted is due to their conformity to the transmissibility of the discipline which are themselves the results of decisions and thus timeless. This will be particularly apparent when we reconsider the idea of intervention in terms of the axiom of choice.

Axiom of choice (Meditation Twenty-two)

As the chapter on Pascal is mainly an illustration, and more than that an illustration of a concept which is, as we can see, incomplete, we will move rapidly on to Meditation Twenty-two which is committed to explaining the axiom of choice. If you recall, Badiou adopts the ZF+C form of set theory which is composed of nine axioms of which choice for some mathematicians is not accepted, leaving them with eight. The meditation commences, therefore, with a consideration of the turbulent history of the axiom of choice. In relation to the axiom of foundation, we noted that it was late to be adopted because it didn't contribute much to the day to day work of mathematicians. Similarly, the reception of the axiom of choice has been problematic, albeit eventually accepted by all set theorists except intuitionists because of the central processes that it allowed mathematicians to perform, even if as an axiom it was highly paradoxical to say the least. This axiom represents for Badiou the interventional form not just in terms of the nature of the axiom, but also in that set theory had to choose to accept it if it wanted to proceed, even if it could not solve the paradoxes central to said axiom when applied to the infinite.[25] Norris does such a good job on these issues that we will leave this discussion and move on to what the axiom actually states.[26]

Badiou defines choice as 'given a multiple of multiples, there *exists* a multiple composed of *a* "representative" of each non-void multiple whose presentation is assured by the first multiple. In other words, one can "choose" an element from each of the multiples which make up a multiple, and one can "gather together" these chosen elements: the multiple obtained in such a manner is consistent, which is to say it exists' (BE 224). In that you can form a set of any set of multiples, this axiom needs some further clarification. Yes, you can form a set out of the elements of multiples that form a set; that is one of the most basic functions of set theory. Badiou says himself that as regards finite sets the axiom of choice 'can be shown by recurrence: one establishes that the function of choice *exists* within the framework of the Ideas of the multiple that have already been presented. There is thus no need of a supplementary Idea (of an axiom) to guarantee its being' (BE 224–5). That said, two features of the axiom of choice should be considered. The first is that each element chosen is a representative of the multiple it is chosen from. In other words, it stands in for all the other elements. Badiou calls it a delegation and gives examples from the system of voting and constituencies. So, the chosen element is like an MP who belongs to a constituency, and also stands in for the constituency. The second is that the axiom of choice determines the efficacy of a function, it does not name a type of set or multiple. If you make such a set of chosen elements, this means simply that to belong to this set the elements themselves must have submitted to the function of 'being chosen'.

The axiom of choice is clarified by Hallward and others in relation to Russell's examples of shoes and socks.[27] Given an infinite number of shoes, you can have a rule for choosing only the left shoe as your representative multiple. However, given an infinite number of socks, all identical, you have to choose arbitrarily: no rule can determine your actual choice. Another illustrative example is that of entering a room full of people all of whom you don't know. You know they all have names but you cannot choose which names make up a set as you don't know their actual names. In this example, the axiom of choice refers to the existential axiom of foundation as it presupposes the existence of multiples in a set indifferently, precisely through the power of arbitrary choice: naming the people. As Norris notes, Badiou likes the axiom of choice because it illustrates the way in which an axiom itself has to be first chosen and then proven. Like in the axiom of choice, which presupposes the well-ordered nature of sets or indeed is one of its preconditions, the mathematician presupposes the logical well-ordered nature of all the other axioms and finds that the axiom of choice, which is not provable, is necessary and exists because of these other axioms. In addition, the axiom of choice has an anarchic side to it as it is a non-rule-governed choice. You must choose, that is a rule, but how you choose is entirely indeterminate. As regards these two qualities, the axiom of choice seems illustrative due to its self-reflexive nature: it shows axiomatic reasoning and it accepts a degree of the random in all formal procedures. Yet there is also a more specific reason why the axiom of choice attracts Badiou.

As we saw in terms of finite sets, the axiom of choice is irrelevant as in all finite sets the nature of your choice can be determined using the basic functions of set theory. So the axiom of choice is only necessary in terms of infinite sets, especially as its legitimacy in set theory is because it founds the idea of well-orderedness as regards the relation between infinite sets. Yet for reasons that do not need to detain us, such a selection in terms of infinite sets is impossible to present a rule for. This is precisely because the rules which mean the axiom of choice is not necessary for finite sets, are not rules which hold the same dominion over different types of infinite or trans-finite sets, particularly the rule of passage for every ordinal that proceeds. If this is the case, then why do we even need the axiom? Badiou sums this up. For set theory, accepting the existence of the function of choice is necessary for several essential, indeed historical, processes in maths including set theory. Yet at the general level it is impossible to define said function. This means that the axiom of choice is entirely existential. We have to accept that there must be choice just as we have to accept that there are sets. As existential functions usually don't concern mathematicians and always concern the ontologist, naturally the axiom of choice is a crucial version of the intervention that we have already described. In short, the axiom of choice is an axiom of recurrent collection, but without the rules of sequential connection that you find in finite sets. You can collect but you cannot say how or why. In a sense, all it says is, like the axiom of foundation, there are multiples in infinite sets because you can collect them into sub-sets so in this sense infinite sets are like finite ones, only you cannot be sure of the rule of this collection in terms of succession and connection. The only connection between the representatives of an infinite set is that all the elements collected are multiples of said set. Which is another way of saying that the axiom of choice is dependent on the indifference of the pure existence of multiples and sets. Collections without connections is another way of saying said multiples are indifferent.

Choice is indifferent

Having established the nature and importance of the axiom of indifferent choice, Badiou goes on to list some of the differences between the implied axiom of choice in finite sets and its application to infinite sets. He notes that as in finite sets, the basic law of choice for infinite sets is that given the existence of a multiple, the existence of another multiple is affirmed, except here what is actually affirmed is a function, that of choice. That said, there are many marked differences relevant to our study especially in terms of the other axioms: the connection between the two multiples is explicit, that the set produced is unique (it is *a* set), and given a property, the set of elements which possess this property is a fixed part of the multiple (left-footed shoes for example). As regards the axiom of choice, the function which is asserted is intrinsic, $(\forall\alpha)\ (\exists f)\ [(\beta\in\alpha)\rightarrow f(\beta)\in\beta]$, which means that its connection to the internal structure of the multiple in question could not be made explicit, nor could you show that the function is unique. This ties the multiple [f] to the singularity of the multiple in question only very loosely and you cannot derive from this multiple

> *a* determined function of f... The axiom of choice juxtaposes to the existence of a multiple the possibility of its delegation, without inscribing a rule for this possibility that could be applied to the particular form of the initial multiple. The existence whose universality is affirmed by the axiom is *indistinguishable* insofar as the condition it obeys (choosing representatives) says nothing to us about the 'how' of its realisation. As such, it is an existence *without-one* ... the function f remains suspended from an existence that we do not know how to present (BE 227).

Naturally, Badiou is setting up the axiom to accord with his already presented idea of intervention. Yet we can also see how the axiom of choice extends indifference into the world of representation such as we noted in relation to the evental site and its effects on the world. What we are talking about here, in the axiom, is a means of creating representations of presentations that are not held within the rules of the state-formed representations of set theory when applied to finite sets. This is important as, to my mind, it is the first moment when Badiou harnesses the power of actually infinite sets for his own purposes. Previously, all they did was further stabilize the natural multiplicity of multiplicities right up to the point of an infinite multiplicity of infinity multiplicities. Here, instead, infinite sets of different kinds, cardinal or non-denumerable sets as we have called them, also show how an indifferent representation can exist in said infinity through the axiom of choice. More than this, in that the axiom is central to foundational concepts in set theory, in particular Cantor's quest for the proof of the continuum hypothesis founded on the now-accepted concept of well-orderedness of sets at the infinite register, the power of representation needs this indifferent representation to proceed. The delegates of infinite sets exist as such, as sets of multiples, and they function as representatives, but they are representatives without names or qualities and they represent an unknown constituency of multiples which we know must exist, but which we can never specify. The axiom of choice in set theory is the proof of the indifference of the event in terms of its consequences relative to the world as it is.

Badiou sums this up neatly when he says: 'What is at stake here is a presentability without presentation' (BE 227). This is another very important statement. The first reason is it allows him to assert: '*within ontology, the axiom of choice formalizes the predicates of intervention*' (BE 227). Formalizing the predicates of intervention is possibly the most difficult task Badiou ever sets for himself and I now think it is fair to say he does not come near to this in *Being and Event*. What we can say of these predicates is that they are indifferent to the world such as it is. While these predicates exist, because the multiple exists, and so it has predicates, the quality of these predicates, even the abstract and indifferent 'qualities' of set theory, remains unknown. By definition, ontologically speaking, an element which exists but which cannot be traced relationally as regards how it differs from its nearest neighbours is a totally indifferent element: we know nothing of its basic qualities beyond existence and we cannot define it as different to any other element yet. We may add, because of the singular nature of this multiple, we can be certain it is not the 'same' as the other presented multiples.

The second reason I tarry here is that up to this point, from our perspective, Badiou has never given a convincing reason for assuming that singularities exist as regards the terms of set theory. Just because multiples are defined in terms of first belonging and then inclusion, and that is all that one can say existentially of them, it does not automatically follow that sets exist which belong and yet which are not included. I would argue that on the contrary the basic idea of a set as a collection of multiples means that no multiple exists that cannot be collected. This is the only existential definition of a multiple that we have. Further, that no set exists that is not a mode of collection which is the inclusion of elements which belong. If this were not the case, then collections would not be collections. Sub-sets depend on belonging, but defining a multiple in terms of belonging, remember in set theory and Badiou's reading inclusion is just an effect of belonging, means that said multiple must be decomposable unless it is the empty set. Multiples are elements which belong so that they can be included. I can see no way here for justifying the existence of singularities or indeed excrescences through the fundamental difference of belonging and inclusion as this is a co-dependant difference wherein if you speak only of belonging or only of inclusion you say nothing that is provable about multiples and sets. Simply put, there are no states in ontology and so no singularities or excrescences either. So, here, the fact that there is a clear axiom for multiples which are presentable but which are not presented, means that it is the case that singularities exist within the transmissible world of set theory.

That said, it must be conceded that what we are actually speaking of is presentability, the fact that a multiple can be presented. This means that in reality, again, we are speaking of an abstract, neutral and thus indifferent 'quality'. All multiples, to be multiples, are presentable. All multiples, to be multiples, in a finite set are also presented. There are no singular multiples in that world. So, when we speak of presentability, we transpose the basic existential quality of all multiples, that they are multiples because they belong to such a degree as they can also be included, to the world where this cannot be constructively proven to be the case. This is the main reason why intuitionist thinking refuses the axiom of choice. There is no 'human' way that the selections of elements from an infinite set can actually be chosen, we just have to assume that based on the formal laws of that set, they potentially could be. They exist but we cannot show that extrinsically. We only present the intrinsic conditions of their existence. Here, then, is a clear difference between the existential assertion of the multiple in the finite world, and the existential assertion of representative multiples in the infinite. These multiples have no determined place, no ordinality, nor any numerable number, no cardinality. We can say they exist but only indifferently. The axiom of choice, then, is both the first time that Badiou justifies his assertion that singular multiples exist, and at the same time proves that said multiples are the only example of representation that is indifferent.

To achieve this, Badiou admits that the axiom of choice allows us to think 'intervention *in its being*; that is without the event' because the event has no place in ontology. So, to make a singularity thinkable, you need to subtract the event. Singularities can be proven to a degree, but their relation to the event cannot. This is not a weakness of the axiom for Badiou but its real power. The inability to decide if the event belongs to a situation 'leaves a trace in the ontological Idea in which the intervention-being is inscribed: a trace which is precisely the unassignable or quasi-non-one character of the function of choice' (BE 227). So, it is because one cannot 'decide' in the axiom of choice, can say for certain what is presented only say that some thing is presentable, that said axiom requires a decision. This seems paradoxical in the extreme but is perhaps just to do with the abuses of ordinary language around such words as choice and decision.

It is time to differentiate the different levels of singularity, particularity, generality and so on in play in terms of these considerations. Already Badiou has named the singularity of choice as a quasi-non-one and so we should begin by defining what a non-one is. A non-one is not the void. Nor is it a set, sets are collections of ones, because a set is just a multiple and counts as one. Indeed, all elements in ontology aside from the void count as one, even the infinite, so non-one has nothing to do with numerability of this order. In that the intervention occurs due to the evental site, the non-one of the event is clearly its being the Two. But the non-one could also refer to the assumed generality of singular multiples in that we say that they exist in general, but we cannot say that they exist in particular. Yet the existential, 'there are some multiples in general', belongs to ontology not to singularity. Plus, singularity and generality clearly do not sit together. In fact, as Badiou has just said, intervention suspends the generality of the being of singularity. It is not a multiple in general because it is not like any other multiple. The void in question here is the fact that a singularity can be asserted indifferently without it becoming general, even though it is in-constructible. This teaches us an important lesson as regards indifference. While it would appear to tend towards generality, indifference is by definition, according to Badiou, entirely singular.

Thus the event, due to choice, presents a void in ontology. It says: no ontology without the axiom of choice, actually no set theoretical ontology without the axiom of choice, no possibility of constructing examples of the axiom from within ontology. So that while mathematical events due to interventions of this sort are good examples of the event, this particular example is not simply exemplary, it is meta-paradigmatic as it presents to us, for the first time, a formal function that shows that singular elements, multiples which belong but which are not represented, exist, that is, they can be represented but only in a non-relational, totally indifferent fashion. For me this solves the earlier problem of the dependency of events on other events, indeed removes any specific dependency. There is, in fact, no talk of circularity here because choice cuts open the circle. Nor is there a need for temporality. Yes, the example of the adoption of the axiom of choice is historical if taken as an example, but if taken as an axiom it is trans-historical: it just shows axiomatically the existence of indifferent multiples in the field of representation with immediate, retroactive effect.

Due to choice, singularities exist and they are indifferent

Summing up, Badiou notes three qualities for the assertion of the existence of the function of choice dependant on one simple fact: 'the assertion of the existence of the function of choice is not accompanied by any procedure which allows, in general, the actual exhibition of one such function, what is at stake is a declaration of the existence of representatives – a delegation – without any law of representation' (BE 229). In other words, that each function of choice is without quality and without generality. Each singularity must be decided on its own terms. The first issue then is that the function of choice is illegal as regards the laws as to how a multiple can be declared to exist. The second is that what is chosen remains anonymous or unnameable. This is because '*There is* a representative, but it is impossible to know which one it is; to the point that this representative has no other identity than that of having to represent the multiple to which it belongs' (BE 229). This is another version of the linguistic function of pure communicability as such: said object is available to name because it exists, but it is an example of the pure intention to signify which means it is an anonymous name. In fact, its name is "to belong to the multiple β and to be indiscriminately selected by *f*" (BE 229).

The anonymity of the axiom then leads to our third quality: it is indifferentiable. For example, the common name of the axiom given above means I can say that the representative is 'put into circulation within the situation' (BE 229), allowing me to say it exists,

> [b]ut I cannot, in general, *designate* a single one of these representatives; the result being that the delegation itself is a multiple with indistinct contours. In particular, determining how it *differs* from another multiple (by the axiom of

> extensionality) is essentially impracticable, because I would have to isolate at least one element which did not figure in the other multiple and I have no guarantee of success in such an enterprise (BE 230).

A better proof of the indifference of singularity one could not ask for. At the same time, Badiou also uses the axiom of extensionality, which was central, you will recall, to our initial presentation of the indifferent contours of set theory's concept of the multiple as being quality-neutral and definable only relationally, to show here another indifference. If extensionality proves the indifference of the pure multiple as such due to relationality, in-extensionality proves the indifference of singular multiples. As we cannot say what elements belong to the multiple, we cannot say which can be represented extensionally. This means, and we are speculating to a degree here, that indifference does not operate under the laws of classical logic. If something is indifferent because it is extensional, if it is in-extensional that does not mean that it is not indifferent.[28]

As the axiom of choice is illegal, anonymous and indifferentiable, Badiou has no trouble saying that said axiom is another name for intervention, on our account a much better name than intervention. More specifically, he shows that the combination of illegality and anonymity creates the indifferentiability of intervention. Due to the limits on the axiom of choice, the only thing it can, in truth, guarantee is 'a form-multiple: that of a function whose *existence*, despite being proclaimed, is generally not realized in any *existent*. The axiom of choice tells us: "there are some interventions"' (BE 230); what he also calls an empty stylization. The formula is fairly simple then. In terms of the axiom of choice, the combination of its illegality and its anonymity results in it being allowed to be represented but only indifferently.

The last point of this quite remarkable meditation is to show how the illegality of choice actually gives rise to the very structure which bans it from its presence. Badiou says 'The consequence of this "empty" stylization of the being of intervention is that . . . the ultimate effect of this axiom in which anonymity and illegality give rise to the appearance of the greatest disorder . . . is *the very height of order* . . . The axiom of choice is actually required to establish that every multiplicity allows itself to be well-ordered' (230). What this entails is that while interventions occur in history, the form-multiple of any intervention is 'a-historical', always to be reclaimed by natural being through the reinstatement of order. Interventions may occur in history, but the being of the intervention, its indifferent and anonymous name, is timeless. Badiou in fact closes on these thoughts. He says that the 'profound lesson' we can learn from the axiom of choice is that history and novelty occur as a result of the combination of the undecidable event and the interventional decision. As regards the being of intervention, he names it 'ineffective' because it 'ultimately functions in the service of order'. This means that intervention as such does not result in disorder, quite the contrary, instead its efficacy comes from the way in which it 'requires rather the initial deregulation, the initial disfunctioning of the count which is the paradoxical evental multiple . . .' (BE 231). It is hard at this stage to say for certain which element, the ineffective or the effective, the indifference of the axiom serves as regards intervention. Everything about the axiom is indifferent, yet the effects of the axiom are not indifferent. There are no neutral events, remember. The being of intervention is indifferent, but the being of indifference is ultimately in the service of order which is, as we saw, differentiated. The question comes down to whether the deregulation of the first effect of the axiom is indifferent. In that Badiou often calls this the void, we could simply say yes. But as he also determines this void as regards the foundationality of the Two, then this seems to disallow it being indifferent. What we can say for certain is that whilst we are now fully in the territory of the event, there is still an essential role for indifference in determining the existence of events. This is the great achievement of Badiou in Meditation Twenty-two, first that singularities exist, and second that they are indifferent.

Fidelity, connection (Meditation Twenty-three)

> I call *fidelity* the set of procedures which discern, within a situation, those multiples whose existence depends upon the introduction into circulation (under the supernumerary name conferred by intervention) of an evental multiple. In sum, a fidelity is the apparatus which separates out, within the set of presented multiples, those which depend upon an event. To be faithful is to gather together and distinguish the becoming legal of a chance (BE 232).

Fidelity is the process by which the consequences of an intervention create a set. It is, in effect, a simple act of judgement. For each presented multiple the faithful subject says 'yes' or 'no' as regards its connection to the name of the intervention and builds up connections of the yeses. Thus, a state of the event is built up, step by step or point by point as Badiou likes to say. This state resembles in many instances actual states, but as you might imagine there are some differences. The process of connection is effectively infinite in as much as there is no known limit to it, but at any time you can halt the process and temporarily define the limits of this evental state because it is strictly sequential. This is how the event becomes a process of consequences, Badiou's real definition, as we saw, of what the event should be considered as. Not as something unique which happens but as a set of consequences due to that occurrence.

There are three named rules for fidelity. First, it is always particular to an event. Fidelity is not some kind of abstract value or general disposition. Second, it is not a term-multiple of the situation but names an operation, a structure, to be evaluated by its results. In this way, it resembles the void from which it emanates because in terms of existence it *is-not*. What exists are the groupings fidelity creates. Third, as it discerns and groups, it counts parts of a situation so it is a kind of state. Badiou calls it variously counter-state and sub-state. This means, perhaps worryingly, that fidelity is institutional. We must be clear, at this stage, in saying that fidelity concerns inclusions, not belongings, so it is not ontological but ultimately concerns the logic of appearing. It may be worth now looking again at the vocabulary of Badiou's definition of fidelity. The first word to think about is 'discerns'. Discernment is the first operation of a fidelity. It spots multiples. The next is 'separates'. Fidelity divides those multiples which are evental from those which are not. The third and final is the word 'gathers', although in general Badiou uses groups. Once a faithful subject can see some multiples, and has decided which are yes in relation to the event, it then groups them all together as one single set. This is basically what fidelity names as a process.

Badiou makes a few qualifications. Although fidelity is specific to a situation, not a general state, it has a universal form.[29] It concerns the connections or non-connections of presented multiples to the $[e_x]$ of the event. Also, there are different levels of fidelity in relation to the same event. He mentions Stalinists and Trotskyists both responding to the event of October 1917, and yet turning on each other. He talks of the century-long conflict between intuitionists and set theory axiomaticians as regards their fidelity to the paradoxes of maths that came to light at the beginning of the twentieth century. And of the means by which the break-up of traditional tonality gave birth to both serialists and the neo-classicists who opposed them. These illustrations show a clear separation between the event as such and its various consequences which we get a better sense of in *Logics*.

Badiou summarizes fidelity in terms of three elements: a situation, a particular multiple and the rule of connection (yes or no). He assigns to this tripartite process a symbol: □. This symbol represents a chain of positive or negative atoms as regards the evental situation. Although fidelity is-not, by which is meant it is not consistent and cannot be counted as one like the multiple itself, at any point, as I said, you can form it into a closed-form state. Any such summary is nothing more than an approximation. That said, you can name the being of a fidelity as made up of the multiple of multiples that have been discerned, according to its operation of connection, dependant on the being from which it proceeds. This leads Badiou to differentiate between the being of fidelity which is finite because it is a finite element of the state due to its being a representation, and the operation of fidelity which is 'an infinite procedure adjacent to presentation' (BE 235). This means that 'a fidelity is always in a non-existent excess over its being. Beneath itself, it exists; beyond itself, it inexists' (BE 236). Due to this, fidelity can always be thought of in two ways, either as regards its being the almost-nothing of the state, or the 'quasi-everything' of the situation. The idea of the quasi-everything in particular looks ahead to the revision of the site of the event in *Logics of Worlds* where an object minimum (a statist almost-nothing) comes to stand in maximally for a new world (a situational everything). This is how self-predication is represented in the category theory of appearing.

Badiou proceeds to consider the relation between the event and the intervention as regards not the one of the state but the two of the ultra-one of the event. Thus the ultra-one of the event is resolved in fidelity as regards the presence of the two. Indeed, fidelity is strictly dialectical concerned as it is with connections as a result of saying yes or no. More than this the duality of fidelity, 'the one-finite' of presentation and the 'infinity of virtual presentation' matches the two sides of the ultra-one of the event. The duality of the event is directly mappable onto its consequences in the world. This means that in the end calling fidelity a state of any sort is incorrect; a mere tactical simplification. Fidelity can never become a state, because it always surpasses itself. Fidelity is an inexistent, ongoing procedure to which all presented multiples are available according to the criteria of □: saying yes or no, being positive or negative. At a more fundamental level, the state is concerned with the count dependant on fundamental ontological operations: belonging and inclusion. In contrast fidelity is concerned with connections of presented multiples to a particular multiple, the event. This means that 'the operator □ has no *a priori* tie to belonging or inclusion. It is, itself, *sui generis*: particular to the fidelity, and by consequence attached to the evental singularity' (BE 237). One can see here in retrospect how fidelity will not easily be captured by set theory and how the introduction of category theory will be the more powerful tool as regards the consequences of an event. As he says in *Mathematics of the Transcendental* 'The ontology prescribed by set theory determines being *qua* being as pure multiplicity "without one" . . .

The ontology prescribed by category theory determines being as act (or relation, or movement)' (MT 13).[30]

Presaging the use of category theory in the second volume Badiou accepts that fidelity can be presented in a typology of proximity to belonging and inclusion. If very close it is more statist. In fact, the only state of belonging relevant to the event is that the event belongs to itself by virtue of the matheme e_x, something we know is absolutely disallowed in ontology. This can lead to different modes of thinking about the event. At one extreme Badiou mentions the thesis that you can only take part in an event if you were 'there'. On the other, the idea that all multiples are dependent on the event. The spontaneist versus the dogmatic stances, as he calls them, are both still statist but to different degrees. So the distance of fidelity is not the distance or not from the event, but the distance of fidelity from the state: 'A real fidelity establishes dependencies which for the state are without concept, and it splits ... the situation in two, because it also discerns a mass of multiples which are indifferent to the event' (BE 237). Real fidelity assesses those multiples the state ignores, and those which are indifferent to the event. It is absolute in its connective relationality: either a multiple is connected to the event, is presented in a situation due to the event, or it is not. Having ascertained that fidelity cannot be a state, Badiou then says but you can still think of it as a counter-state: 'what it does is organise, *within* the situation, another legitimacy of inclusions. It builds, according to the infinite becoming of the finite and provisional results, a kind of *other* situation, obtained by the division in two of the primitive situation' (BE 238).

Badiou concludes on fidelity as follows: 'One of the most profound questions of philosophy ... is that of knowing in what measure the evental constitution itself ... *prescribes* the type of connection by which a fidelity is regulated ... Is it possible that the very nature of the site influences fidelity to events pinned to its central void?' (BE 238). We already know that some mainstream opinion says yes to this just as it says yes to the possibility of a specific event being the basis for the next event of that order. One revolution gives birth to another and so on. It is also clear that one of the central axioms of our study is the absolute non-relationality of the event due to the void. In that the event is indifferent in its being, and different in its effects, there can be no difference in kind as regards events not in terms of the ontological basis of paradox from which a site is formed from said event. I go further. If the event is an example of real change, that is an absolute difference from all that currently exists, it must originate in a truly non-relational, indifferent event.[31] For now Badiou side-lines this very thorny issue (one can feel here that he wants events to give rise to events and he wants certain sites to determine certain types of fidelity but he is wise enough to realize that his ontology does not allow for this). Instead, he says that if we accept, as I am insisting, that there is '*no relation* between intervention and fidelity, we will have to admit that the operator of connection in fact emerges *as a second event*' (BE 239). He speaks of this as a 'complete hiatus between e_x, circulated in the situation by the intervention, and the faithful discernment of what is connected to it' (BE 239). We now know that whether or not we call fidelity an event, the means by which you access the event, using indifferent naming, is radically different from that which you need to access its consequences so that □ basically marks the difference from naming in set theory to relation in category theory. He calls this other event the 'operator of fidelity', but we can now call it category theory as the basis for the logic of appearing and refer the reader to our second volume for a full explanation of what we mean by this designation.

Taking this to be the case the final sentence of the twenty-third meditation is redolent to say the least: 'the more distant the operator of connection □ is from the grand ontological liaisons, the more it acts as an innovation . . .' (BE 239). The true power of the event is not in terms of its ontological paradox but as regards the logic of its appearing. We will never know the event from the perspective and language of *Being and Event* itself, we need *Logics of Worlds*. In that the subject is recomposed in the second volume in terms of a body that is made up of points, not an intervention that is then determined by fidelity, we have said enough on this early version of subjective formalism and must wait for the second volume to give us what appears to be the final word on the matter.

7

The Generic

Continuum hypothesis (Meditation Twenty-seven)

Part VI of *Being and Event*, 'Quantity and Knowledge', is primarily concerned with maintaining the event against the victories of constructivist thought in the twentieth century. Constructivism is by far the most dominant mode of thinking of our age, spanning Anglo-Saxon logic, ordinary language theory, Foucauldian discourse and Deleuzian assemblages to name but a few entries into its impressive, multi-volume definition. Most prominent for Badiou amongst these is Gödel's constructible set theory.[1] Kurt Gödel, in the 1930s, was able to show that Cantor's pursuit of well-orderedness and the continuum hypothesis are provably impossible, yet at the same time if you accept that the universe is constructible rather than real, this really does not matter. We have mentioned Cantor's continuum hypothesis several times and now it is perhaps the hour to better explain what it states and its implications. To do this we will rely on Hallward's excellent treatment of the issue backed-up by strong scholarship across the board.[2] If you recall, Cantor, having proven that ω_0 exists, the first actually infinite set of natural numbers, proceeds to demonstrate that there must be other actual infinities of different sizes. For example, there must be an infinitely larger set of real numbers that includes irrational and transcendental numbers, numbers particularly useful for the physical sciences as they allow us to capture motion. These numbers are usually defined as the geometric continuum of points on a line. Cantor found a way to determine what he called c, the infinite set of all real numbers, which is exponentially larger than ω_0, here called $\aleph_0$, written as $2^{\aleph 0}$. There is a clear order of continuity between $\aleph_0$, the set c, $2^{\aleph 0}$, and the next largest set, $2^{2\aleph 0}$. If Cantor can show a direct continuum between the sequence of sets of infinite natural numbers, which as we saw was ordered due to the n + 1 sequence as $\aleph_0$, $\aleph_1$, $\aleph_2$ and so on, and those of real numbers, the implications for science and mathematics would be immense. Cantor believed that there was a direct relation by stating that $\aleph_1$ was the same size as $2^{\aleph 0}$. That said, he was never able to prove it and eventually Gödel and Cohen were, in the 1960s, finally able to show that it was not possible to prove CH, as it came to be called, from within the axioms of set theory itself.

This may seem a tad technical but the implications for our study as regards the provability or not of CH are significant. If CH were proven, then there would be a direct link between 'physical continuity and number', in addition to which 'everything within the transfinite universe could be thought of as in its appropriate place'.[3] If it is not provably true, and in the end it was not, then there is an absolute break in the universe of transfinite numbers between $\aleph_0$ and the power set of all the parts of $\aleph_0$ or $2^{\aleph 0}$. As Hallward says, 'A universe that denies CH would thus accept a constituent degree of ontological anarchy. It would tolerate the existence of sets that could not be assigned any clear place in an order that would include them'.[4] This realization is the basis of the work done by Cohen and Easton that lead to the complete formalization of the axiom of choice, the role of the generic and the necessity of forcing. It is also the main reason why Badiou rejects constructible theories such as those of Gödel, which seek to limit the impact of the immeasurable by excluding it from any constructible world. Badiou addresses CH in Meditation Twenty-seven where he clearly differentiates between the continuous and the denumerable as concerning the 'pure errant principle' versus 'the closing down or blocking of this errancy' (BE 281).

Badiou recasts the entire history of Western philosophy along this fault-line between errancy and denumerability or how a state deals with the excess of its parts over the whole due to the void. He negatively describes three archetypal approaches to this dialectic, before favouring a fourth. The first of these Badiou names Grammarian or programmatic. It states that language is the un-measure. Due to this, the state then determines what can be named due to the already existent parts of a situation of language, and leaves everything else as the unnameable and thus unformed. This first position is clearly intolerant of indifference and entirely dependent on an ontology of mutual differentiation of beings. Badiou names it as the core of Gödel's idea of constructible sets, founded on Leibniz's philosophy of indiscernibles. It is this Grammarian approach that the sixth part of *Being and Event* and a large portion of *Logics of Worlds* tries to undermine.

The second 'endeavour' works in the opposite direction. In this instance, it is precisely indifference as such which forms the basis of what is: 'the excess of the state is only thinkable because discernment of parts is required. What is opposed this time, via the deployment of a doctrine of indiscernibles, is a demonstration that it is the latter which make up the essential field in which the state operates ... that any authentic thought must forge for itself the means to apprehend the indeterminate, the undifferentiated, and the multiply-similar' (BE 283). He speaks of Cohen's generic sets in this regard and of the political theory of Rousseau, but he may as well speak of our own study.[5] In that the three great orientations of thought are set to be usurped by a fourth, associated with Marx and Freud and the valorization of the event as intervention on thought, and it is clear Badiou holds to this fourth position, we must submit to the fact that his is not a thought of indifference and yet the more he speaks of this second great endeavour, the more we have to conclude this does accurately describe Badiou's ontology while failing to describe any notable grouping in the history of philosophy which has never, until the work of Agamben, Badiou and Laruelle, consisted of a group of thinkers committed to indiscernibility due to indifference.

For example, he says of this second grouping that its take on representation is 'what it numbers without discerning'. That it works rationally to create a 'matheme of the indiscernible, which brings forth in thought the innumerable parts that cannot be named as separate from the crowd of those which – in the myopic eyes of language – are absolutely identical to them' (BE 283). From such a perspective, he says the mystery of excess is not reduced, as is the case for constructible thinking. Rather, 'Its origin will be known, which is that the anonymity of parts is necessarily beyond the distinction of belongings' (BE 283). It is perhaps only the last statement that helps us realize that rare though this disposition is, still Badiou is not a thinker of indifference but a thinker because of indifference. He would never countenance a realm beyond belonging wherein indifferential multiples exist in number. Rather, all that he will countenance is that indifference will allow for the presentation of something which belongs but is never included. If said element never belonged, then it would enter into something like a field of infinite, undifferentiated multiples à la Deleuze, but the whole point of the indifference of being is how being comes to be operative on existence due to the specific indifference of its operations, not because being is simply an indifferent infinity of multiples yet to be realized.

The third endeavour is perhaps the most familiar, that of transcendentalism. You set a stopping point for errancy by suggesting that it results from the failure of language or the undifferentiated to fully appreciate the sheer size of presentations. To remedy this, you 'differentiate a gigantic infinity which prescribes the hierarchical disposition in which nothing will be able to err any more' (BE 284). Badiou mentions the set theory doctrine of large cardinals in this regard, and of course 'all of classical metaphysics' (BE 284). We have already mentioned that these three dispositions are set up to be distinct from the fourth which is the thought of the event which Badiou historicizes due to Marx and Freud. This simply says: do not be afraid of 'the un-binding of being, because it is the undecidable occurrence of a supernumerary non-being that every truth procedure originates' (BE 284–5). Leaving aside the problematics of suggesting two thinkers, Marx and Freud, so tied to the philosophy of difference, could ever represent a philosophy of the event due to indifference, it is obvious that the main distinction between endeavour two and four is that endeavour two, in maintaining that indifference is the norm, cannot admit to a theory of its being evental. That said, from the opposite perspective, the big problem with the way Badiou presents the event here is that it becomes tantamount to a dialectic that is intrinsically a feature of philosophical difference and totally absent from any thinking of indifference: 'It states, this fourth way, that on the underside of ontology, against being, solely discernible from the latter point by point (because, globally they are incorporated, one in the other, like the surface of a Möbius strip), the unpresented procedure of the true takes place . . .' (BE 285). While one can concede that disposition two disallows the event and that, as we saw, that which allows for a philosophy of indifference, set theory, also demands that there be an intervention, axiom of choice; the weakness of this final statement is not the preferred alternative. The combination of Marx, Freud, undersides, one in the other, and Möbius strip show that while disposition two is incomplete because it cannot allow the discernibility of the event into the indiscernible nature of indifference, equally position four fails because in presenting the event in the terms of a standard, dialectical, neo-Hegelian language of differentiation, it leaves the centrality of indifference to the event, totally indiscernible.

Returning to the point at hand, CH is a central moment in the history of mathematics between those, like Cohen and Easton, who accept errancy and axiomatize it, and those such as Gödel who do not accept errancy and thus exclude it from a constructible theory of sets. So we see that in contrast to the ZF+C nine axiom set theory that Badiou favours, Gödel was able to use a more elegant seven axiom version of set theory to prove that every situation is constructibly consistent. To do this he had no need of the two axioms that, philosophically and politically, are of greatest importance to Badiou, those of foundation and choice. Badiou's argument with Gödel's constructivism is that without foundation and choice yes, ours is a totally constructible universe and he commits almost the totality of *Logics of Worlds* to show this to be the case. Yet at the same time without foundation (the real), and choice, the (event), such a universe is boring. So boring indeed that mathematics is only able to live in it for a handful of decades before in the 1960s it starts to break out of constructible set theory by developing from axioms such as foundation and choice. So that while Gödel is almost able to totally negate excess and errancy in every situation, a simple fact plus a basic attitude means that ultimately he fails. The fact is that belonging and inclusion are not the same, meaning the recount of the state is always in excess of the initial situation and this can never be forgotten. The attitude is simply, super-stable states disallow innovative thought and eventually, if mathematicians were able to prove a constructible universe, they were also the first to respond to its limitations.

To conclude, constructivism has one basic axiom which is that everything is constructible in language except the fact of that language itself (Gödel's incompleteness theorems). Derrida is a constructivist for example on this reading. Badiou's position is this is true except that there are two non-constructible elements, the void which is real, and the event, which you have to choose. Neither can be proven constructibly, but both can be retroactively shown to be the case due to the deductive conditions, the language, of all constructivist thought. You don't have to accept the real of the void and the choice of events, but if you are a constructivist you cannot disallow them within your language, and if you are a militant of any order, you will desire the effects of allowing them eventually.[6]

This is as much time as we need to commit to the central heart of the sixth section. *Logics of Worlds* is a profound study of constructible thought relative to states or worlds so I see no need to spend any more time on constructivist thought itself here. While all the other elements of the sixth section are covered elsewhere in the study. The difference between quantity and knowledge will be central to the process of interventions and fidelity. While the important section on Easton's Theorem (1970) and its development of the conceptless choice will be considered towards the end of this book in light of his conception of choice and the generic. The role of the discernible and Leibniz's theorem of indiscernibles is of central interest to Badiou and our own study, and we refer to it numerous times so there is no need for a specific section on it. We will not, therefore, dedicate a chapter to section VI, for no other reason than all of its concepts are developed in more detail elsewhere. Instead, we will commit our increasingly limited spatial resources to the remarkable Part VII and its foundational comments on what Badiou, after Cohen, calls the generic.

The thought of the generic (Meditation Thirty-one)

The generic and its relation to forcing are the last two major concepts inherited from set theory pertinent to the event, truth and the subject. In particular, the generic is such that it 'will found the very being of any truth' (BE 327). While the detail of the generic is not given until Meditations Thirty-three and Thirty-four, the fact that Badiou says that the terms generic and indiscernible are 'almost equivalent' (BE 327) means we are already well-versed in many aspects of the function. In addition, in that our thesis is that indiscernibility is tantamount to a mode of indifference, in this instance the indifference of quantity, it should be clear first that the concept of the generic will found our theory of truth in relation to indifference, and second, insist on the foundation of truth in our formulations of indifference. Badiou explains specifically that he prefers the term generic in this instance because it does not have the negative connotations of indiscernibility 'which indicates, uniquely, via non-discernibility, that what is at stake is subtracted from knowledge' (BE 327). In contrast, by his reckoning, the generic 'positively designates that what does not allows itself to be discerned is in reality the general truth of a situation, the truth of its being, as considered as the foundation of all knowledge to come' (BE 327).[7] Apart from being positive the generic also contains the idea of generality: the general truth of a situation for a knowledge to come. This is important as 'everything is at stake in the thought of the truth/knowledge couple' or 'thinking the relation – which is rather a non-relation – between, on the one hand, a post-evental fidelity, and on the other, a fixed state of knowledge' (BE 327). This latter state is what Badiou habitually calls, with a negative register, the encyclopaedia.

Badiou wants a means by which the procedure of fidelity crosses already existing knowledge, starting as the name of the event. The result is a five-part schema for the event as consequences to knowledge by means of the procedures of fidelity.

1. Study local forms of a procedure to fidelity. These he calls enquiries.
2. Present a distinction between the true and the merely veridical (truthful within knowledge). This should lead one to say that every truth is infinite.
3. Analyse the question of generic truths.[8]
4. Consider the way in which a procedure of truth removes itself from the jurisdiction of knowledge. Badiou calls this avoidance.
5. Finally, present 'the definition of a generic procedure of fidelity' (BE 328).

This is the program he sets up for the considerations into the generic which are to follow so that he can establish one of the foundational separations of his entire work, that between the true (fidelity to events) and the veridical (constructible worlds).[9]

Discernment and classification

Badiou defines knowledge in terms of two procedures: discernment and classification. While classification is the means of grouping together multiples due to the properties they hold in common, discernment determines presentation of multiples to the world of knowledge, and as you may have realized, classification concerns the representation of these multiples in relation to conglomerations of states. The encyclopaedia, as Badiou calls it, first realizes the summation of all judgements pertaining to what can be discerned, and second works to classify these parts due to property similarity and already existent categories of classification. These two functions are inter-related of course, presenting a very real sense in which the world of discourse is entirely relational and in which 'events' as the encyclopaedia encounters them are always pre-prepared by other already existing 'events'. As regards the event as such, it is unnameable within the encyclopaedia not because it cannot be named or its name is somehow terrible and unacceptable, but because the event is simply never presented to the encyclopaedia: it cannot be discerned and thus it cannot be grouped, even though it is present. This means presentation due to discernment and pure presentation as such, generic presentation, are radically dissimilar. Although we have reservations as regards this dialectic, we can take the non-presentation of the event to a situation as a historic quality, while its being unnameable is simply an issue of signification. As we know, language can signify anything that it can discern, so that when it names something as an event, or as without name, this is perfectly representable for it. What it cannot tolerate is something that presents itself as indiscernible, and the only way this can happen is if it is indiscernible to language in the first instance.

The 'person' who commits themselves to a fidelity Badiou calls a militant and their power has nothing to do with what they know, only with what they can connect to what they believe to be an event. This results in Badiou's basic logic of the event. A militant tests multiples to see if they are connected to the event. If they are she designates them x (+), if not x (−). Subsequently a 'report' is generated as regards chains of positive connections and chains of negative. This is what is meant by an *enquiry*, 'any finite set of such minimal reports' (BE 330). So a process has 'militated' around a series of encountered multiples, and it has designated connections or non-connections of each to the event. It is apparent already that an enquiry resembles in many aspects the encyclopaedia itself as it combines 'the one of discernment with the several of classification' (BE 331).

Truth and knowledge: the indifference of avoidance

It is clearly incumbent upon us that we have to ask after the dialectic between truth and knowledge. Every group of finite multiples is nameable in a constructible universe. This is as true of the report due to fidelity to the event as any other finite grouping. This means we now need that suggested difference between the merely veridical and the actually true. Badiou says 'we will term *veridical* the following statement . . .: "Such a part of the situation is answerable to such an encyclopaedic determinant"' (BE 332). In contrast, the truth statement of the event is: '"Such a part of the situation groups together multiples connected (or unconnected) to the supernumerary name of the event"' (BE 332). As Badiou famously says in *Manifesto for Philosophy*, a 'truth makes a hole in knowledge' (MP 80)[10] or, as he says in *Theory of the Subject* 'All truth is new' (TS 121).

Badiou uses a mode of classical logic, excluded middle, to say because an encyclopaedia can always count finite groups of multiples, truths must be infinite. This formal means of reasoning solves a problem which is that an enquiry cannot itself discern the true from the veridical, as an enquiry is always a representation and thus veridical in essence: 'its true-result is at the same time already constituted as belonging to a veridical statement'. This produces what Badiou calls the paradox of a multiple which is random, generic, indiscernible and thus indifferent ('subtracted from knowledge'), (BE 332), yet which is always already part of the encyclopaedia. 'It is as though knowledge has the power to efface the event in its supposed effects, counted as one by a fidelity; it trumps the fidelity with a peremptory "already-counted"' (BE 332–3). Badiou goes on to explain that this is only the case when effects are *finite* resulting in the central law: '*the true only has a chance of being distinguishable from the veridical when it is infinite.* A truth (if it exists) must be an infinite part of the situation' (BE 333). So, a truth cannot be determined through direct enquiry, as that would make it veridical. However, in the case of the infinite which is non-denumerable, there is a possible home for truth. Thus, a truth can be generated by directly proving that it is impossible for any finite grouping but that infinite groupings exist, thus so do truths. Here actual infinity works to support the event in a positive manner through the observations we have already made about how trans-finite numbers confound CH's insistence on a well-ordered relation between $\aleph_0$ and $2^{\aleph_0}$ yet do not negate well-orderedness in its entirety thanks to the axiom of choice.

That said, Badiou, while rather easily defining a truth as infinite, appreciates that being infinite is not the same as being true. Knowledge, since Cantor, 'moves easily amongst the infinite classes of multiples which fall under an encyclopaedic determinant' (BE 333). More than this there are other problems to be dealt with as regards the being of truth. In terms of the quality of an enquiry, the finite result is very different from an encyclopaedic determinant because the procedures of their production are non-relational. This means multiples are indistinguishable to an encyclopaedia because, as pure multiples, they all fall under a single determinant which is that of 'being indiscernible'. Yet if we want to differentiate truthful and veridical statements this is not enough. The means of differentiating truth enquiries due to a radical difference of *procedure* (event – intervention – fidelity) proves nothing if the presented multiples are all the same. Accordingly, these multiples must not just be different as regards the procedure of their production, when they are presented they must remain 'indiscernible and unclassifiable for the encyclopaedia' (BE 333). And they must be presented because while they are indiscernible multiples as such, they are not just indifferent to the encyclopaedia, they are indifferent to each other making the event quasi-transcendental when we know all events are immanent to a particular situation: there are no neutral events. Badiou says their being infinite is part of the package of the ontological difference between truth and the veridical, but it is not, for him, enough.

Having established that infinity alone is not sufficient 'to serve as the unique criterion for the indiscernibility of faithful truths' Badiou moves to propose that a better criterion comes in the form of the generic. Specifically, here he is considering the strategy he calls 'avoidance'. Badiou concedes that any finite group must be determined by the encyclopaedia's processes of discernment and classification. Avoidance, on this reading, is a means by which a group of finite, represented elements are able to not be determined in this way. To do this, our group needs to present to the encyclopaedia the simultaneous presence of positive and negative aspects in a manner that is undecidable for knowledge. If you remember, that is all militancy is, deciding on positive and negative connections. The encyclopaedia has no problem with positives or negatives. Further, if you take a group in which x possesses a property and y does not, then the group $\{x,y\}$ is an object of knowledge. However, if this finite part is presented in a generic fashion 'it is indifferent to the property, because one of its terms possesses it, whilst the other does not. Knowledge considers that *this* finite part, taken as a whole, is not apt for discernment via the property' (BE 335). Avoidance due to indifference is still an element of knowledge, but it allows Badiou to think how a finite set of evental elements could be represented, and thus exist in knowledge, and yet be entirely new to knowledge. Badiou's formula for this is as follows:

> Take an enquiry which is such that the terms it reports as positively connected to the event (the finite number of x (+)'s which figure in the enquiry) form a finite part which avoids a determinant of knowledge in the sense of avoidance defined above. Then take a faithful procedure in which this enquiry figures: the infinite total of terms connected positively to the event via that procedure cannot in any manner coincide with the determinant avoided by the x (+)'s of the enquiry in question (BE 336).

Badiou is gambling on the fact that as regards any enquiry after the evental, there will always be within the group of positive connections, at least one element that, as regards their determination by knowledge or language, namely they have this property, *does* not have this property. This element is indifferent to knowledge which cannot see contradiction because it is purely differential, but it is discernible within the set of elements that make up a truth. Please note, these elements are not those which are negatively charged, they too are indifferent in this case as regards the event. Badiou sums up this rather complex situation by saying: 'there are elements in the class which have the property and there are others which do not. This class is therefore not the one that is defined in the language by the classification "all multiples discerned as having this property"' (BE 336–7). It is patently obvious the role indifference has to play here at this crucial moment. The indifference of the presence of this at least one extra element is without property, yet it is present and, as we saw, this means it is indifferent because it is differentially related due to something other than its determined and constructed qualities. While such elements do not exist for knowledge, they do exist for truth. Thus, the whole of Badiou's central differentiation between knowledge (veridical) and truth, and more importantly the rather involved manner in which he establishes that in knowledge something truly new can be presented, essential for proving the event is truly evental, depends entirely on this specific, technical use of property indifference. In reality, this is just a different way of considering the axiom of choice: you know such-and-such parts of a set will possess positive and negative 'charges' but you do not know which does and which does not, at least not yet. The generic then allows one to pursue a set of multiples due to an operational consistency, they are either positives or negatives, yet remain at a quality-neutral or indifferent level, but we don't know which are positive and which are negative. So one can say that these multiples will be well-ordered, but we can only speak of their potential to be so, we cannot name or in any way designate this ordering.

Badiou states that it is sufficient for only one enquiry to avoid the determinant for an infinite faithful procedure to occur. He also adds that this is a reasonable assumption 'because the faithful procedure is random, and in no way predetermined by knowledge. Its origin is the event, of which knowledge knows nothing ... The multiples encountered by the procedure do not depend on any knowledge' (BE 337). Going further, as the enquiry has nothing in itself to do with the determinant there is nothing in knowledge to stop it occurring. This is a rather complex mix of classical logic's use of double negatives, and a more intuitionist liberalism in allowing an enquiry to consist of contradictory elements. Throughout this rather dense justification for truthful enquiries, Badiou is treading a very thin line of reasonable assumptions, as he often calls them, using phrases such as 'there is no reason not to assume' and so on. More than this, as we said, avoidance requires the presence of contradictory elements, does possess quality and does not possess quality. Yet it asks that the same vigour not be extended to its own connections, where negative connections which are enquiry-indifferent, seem to have no impact on the enquiry. Then, as we have just shown, the proof of the probability of quality indifferent yet presented elements is based on a double negative, there is no reason not to assume they exist as nothing about knowledge can stop them. Finally, there is the need for procedures to be infinite and yet doubt over the proof that they actually are. So there is quite a bit to take on trust here before the reader concurs with Badiou's conclusion: 'if an infinite faithful procedure contains at least one finite enquiry which avoids an encyclopaedic determinant, then the infinite positive result of that procedure (the class of x(+)'s) will not coincide with that part of the situation whose knowledge is designated by this determinant' (BE 337). Or, accept the axiom of choice, then observe its effects not on ontology but on what will come to be called the logic of worlds, here the encyclopaedia of knowledge.

Generic procedure

Having stipulated that it is enough for one such enquiry to exist, Badiou speculates as to what a group of such enquiries might resemble. Such a class of multiples will not contain any parts that are 'subsumable under a determinant' and as such will be '*indiscernible and unclassifiable* for knowledge'. This then comes to be the basis for what Badiou calls the generic '*a truth is the infinite positive total—the gathering together of* x (+)*'s—of a procedure of fidelity which, for each and every determinant of the encyclopaedia, contains at least one enquiry which avoids it.* Such a procedure will be said to be *generic*' (BE 338). The fifth section of the meditation details the qualities of this assumed generic procedure. Badiou explains first that the generic is a part of the situation, and second that it is indiscernible to the situation. Badiou moves to ask after this represented one, this part, which is also indiscernible. It has no particular properties so it is simply *a* part in a generic sense. Such an indiscernible inclusion is without property except 'that of referring to *belonging*'. Said part is anonymous as it has no other mark 'apart from arising from presentation', of being 'composed of terms which have nothing in common that could be remarked, save belonging to *this* situation' (BE 339). All that the terms of the situation share is their being: that they belong to the situation. This means that the part 'solely possesses the "properties" of any part whatsoever' and this is why Badiou names it generic: 'because, if one wishes to qualify it, all one cans say is that its elements *are*' (BE 339). What this shows is that the part belongs to what Badiou calls the supreme genre: 'The being of the situation as such' (BE 339). This being the case every indiscernible part is part of the truth of the entire situation 'insofar as the sense of the indiscernible is that of exhibiting as one-multiple the very being of what belongs insofar as it belongs' (BE 339).

If we contrast the indiscernible part with the discernible part of a constructible language we can see that nameable parts, naming is the mode of discernment for knowledge, does not refer to being-in-situation as such but rather to clear and known particularities that language has made knowable. So such a part presents the general truth of presentation as such, 'whilst a determinant of knowledge solely specifies veracities' (BE 339). In that the generic is another name for the indiscernible we are already on track for saying that the generic is indifferent so that when Badiou concludes 'there is no truth apart from the generic' (BE 339), he is effectively saying: there is no truth apart from that which is indifferent: 'Evidently, everything hangs on the possibility of the existence of a generic procedure of fidelity' (BE 339).

But, Badiou asks, do truths actually exist? He thinks he can prove this de facto, by giving brief examples from history in the fields of love, art, science and politics, but this remains rather tenuous. Then he thinks he can also prove them de jure by citing Cohen's 1963 discovery of the generic in relation to set theory. To satisfy ourselves as regards the former, de facto proof is difficult and the reader may have noticed a studious avoidance of Badiou's concept of 'conditions', and his examples 'from history' of his various concepts. In short, our position is, there is no historical proof of truth or event possible. This leaves the de jure proof. For that we now have to plunge into Cohen's twin concepts of the generic and also the means by which one can force a generic situation to come into existence. Only then could it be possible to say if, for example, atonality is or is not an evental truth.[11]

The matheme of the indiscernible (Meditation Thirty-three)

Badiou starts his consideration of Cohen by saying that mathematical ontology has no concept of truth, as truth is always post-evental and as we know the event is prohibited in ontology. That said, the being of a truth, which *is not* the truth, 'exists' as it is in a situation even if it is so as the indiscernible and generic. So, although truth as such is not thinkable, the being of truth relative to a situation is, meaning if ontology could produce the concept of the generic multiple, 'an unnameable, un-constructible, indiscernible multiple' (BE 355), then it could be said that 'there exists an ontological concept of the indiscernible multiple' (BE 355). The result of this reasoning is that said multiple can be called true due to its being indexed to an assumed event. This is a major achievement because 'Consequently, ontology is compatible with the philosophy of truth. It *authorizes* the existence of the result-multiple of the generic procedure suspended from the event, despite it being indiscernible within the situation in which it is inscribed' (BE 355). How do we bring this amazing result about?

The first precondition is that if the indiscernible multiple exists, proving that truth exists due to the event, it only exists against the backdrop of discernibility: 'indiscernibility is necessarily relative to a criterion of the discernible, that is, to a situation and to a language' (BE 356). This is an extremely important stipulation for the history of philosophy. Key thinkers of the possibility or not of singularity after Heidegger, specifically Derrida and Deleuze, struggle with the realization that if singularity 'exists' it must exist by virtue of a language, yet if it exists by virtue of language then one must conclude that language admits to no singularity and thus the singular does not exist. This has resulted in singularities that are corrupted as soon as they are proposed, or that are never truly singular in the first place, and so on. Badiou appears to be saying the same thing here but, as we shall see, Cohen's great innovation is to allow one to speak of the indiscernible from the perspective of the discernible in such a way that it remains indiscernible. Indeed, the very concept of indiscernible is due to the structure of the discernible itself. Thus one can admit to singularity as truly singular yet represented. So far, in Western thought, we have never had access to such a thing.

What Badiou proposes is to take a fully determinate situation which is the basic model of set theory, it is called the 'fundamental quasi-complete situation' or 'ground-model' – the totality of everything that is discernible – and introduce a meaningless sign therein which will stand for the indiscernible part. This sign ♀, is absolutely indeterminate and therefore entirely generic. As ♀ is not nameable, it is not in any sense itself a name of the indiscernible, filling in the absences of what it consists of must occur via procedures that themselves operate within the nameable. So one is using the procedures of discernment to discern something as absolutely indiscernible. One can see here the similarity with what we have called enquiry, but it must be remembered that now we are speaking of pure mathematical ontology which is, in fact, not procedural but structural: the procedures only allow us to see and speak of the structures.

So this is our basic position. We start from supposing that a multiple exists in the initial situation. This multiple, determined ♀ although not named as such, functions in two different ways as regards its being indiscernible. 'On the one-hand, its elements will furnish the substance-multiple of the indiscernible, because the latter will be a *part* of the chosen multiple. On the other hand, these elements will condition the indiscernible in that they will transmit "information" about it' (BE 357). Or, the multiple in question will combine the material for the making of an indiscernible and also determine its intelligibility or what we would call its communicability. We must say as an aside that this view on conditions is much more satisfying than the four generic procedures of love, politics, art and maths that Badiou speaks of more generally.[12] The specific role of conditions is 'that certain groupings of conditions, conditions which are themselves conditioned *in the language of the situation*, will make it possible to think that a multiple which counts these conditions as one is incapable, itself, of being discernible' (BE 357). By this Badiou means not discerned in the original quasi-complete situation. This is the multiple he designates by the sign ♀ with the reservation that it does not belong to the situation and will remain permanently supernumerary, outside the count, even after its conditions have been filled in. 'The idea is then that of seeing what happens if, by force, this indiscernible is "added" or "joined" to the situation' (BE 357). We encounter here for the first time the combination of the generic, ♀, and the process of forcing it on the situation almost to see what happens if one did that. This is yet another example of the retroactive logic typical of Badiou's work in that the supplement of the event, forced on a situation, comes 'after' the intervention.

Badiou is specific in his use of language in these deliberations as he considers the sense of the meaning of adding an indiscernible once it has been conditioned, not constructed or named (both operations of a constructible universe). In particular, what can this mean in reality if you cannot name and thus construct the multiple in question? Badiou suggests that one constructs names from within the situation for every element obtained by the addition of ♀. This implies that we do not know which multiple of *S*(♀), Badiou's name for the situation, is named by each name. Rather what we are saying is that each multiple can be named, in general or generically. As Badiou says, 'we will know that there are names for all' but we will not know what those names specifically are (BE 358). These generic names open up the possibility of thinking properties for the situation *S*(♀), properties which themselves will again be entirely generic. Badiou sums up this odd situation by saying: 'the indiscernible leaves a trace in the form of our incapacity to discern "an" extension obtained on the basis of a "distinct" indiscernible' (BE 358).

It should be apparent that the generic operates very much like the void as regards being as such. The mark of the indiscernible is what it does not allow one to do, talk of the ♀ in terms of specificity and distinctness, but this mark exists and this allows for one to begin to condition ♀ accordingly under the licence of generic, indifferent procedures. As we shall see this means being able to speak about ♀ very specifically in a formal manner, with very strong speculations based on the means by which we know such a name must operate, without ever stipulating what said name actually discerns. The result is a robust formal means of thinking indifferently that is, as far as I am concerned, the first attempt at such a process in the history of thought: an entirely formalist mode of thinking which, however, admits a non-constructed, entirely indifferent universe. For me, at least, this marks a new epoch in the history of philosophy, healing the schism between analytical and continental traditions through a critical insistence that both must now allow themselves to be systematic philosophies of indifference.

Easton's Theorem (Meditation Twenty-six)

The matheme of the indiscernible owes as much to Easton as it does to Cohen, in particular Easton's famous theorem of 1970 which Badiou considers in Meditation Twenty-six.[13] We need to outline this theorem for two reasons. First, because it is the moment when Badiou starts to explain that the relation between a cardinality and the set of its parts is 'shown to be rather an un-relation, insofar as "almost" any relation that is chosen in advance is consistent with the Ideas of the multiple' (BE 279). This is the basis of our previous reconsideration of the entire concept of relationality, and perhaps a radical revision of what non-relationality means, specifically here a relation so neutral or generic that it runs the risk of being called the indifference of indetermination. Second, because Easton's Theorem takes us close to the idea of the 'conceptless choice' as a solution to the threat of the indifference of indetermination which menaced both Hegel and Deleuze in equal measure.

What Easton shows is that we know quite a bit about the 'size' of any cardinality, for example the set of its parts must exceed the size of elements that belong, and one or two other technical elements I won't linger on. From this, Easton posits that while you cannot determine exactly what the size of an infinite or successor cardinal is, you can say certain things about it that are 'coherent with the Ideas of the multiple you choose' (BE 279). Badiou explains that this means '*if* these ideas are coherent amongst themselves (thus, if mathematics is a language in which deductive fidelity is genuinely separative, and thus consistent), *then* they will remain so if you decide that, in your eyes, the multiple $p(\lambda)$ has as its intrinsic size a particular successor cardinal π; – provided that it is superior to λ' (BE 279–80). Such a string of deductions allows Badiou to say of Easton's Theorem that it 'establishes a quasi-total errancy of the excess of the state over the situation. It is as though, between the structure in which the immediacy of belonging is delivered, and the metastructure which counts as one the parts and regulates the inclusions, a chasm opens, whose filling in depends upon a conceptless choice' (BE 280).

The conceptless choice as regards actual quantity of the infinite sets demonstrates one of the great confusions as regards indifferential logic, if such a thing can be said to exist. If one element is determined as different from another purely in terms of relation not quality, then this relation must be a quantifiable one. If this quantifiable difference is all we have in set theory, then it is hard to see this as actually indifferent. Certainly the difference between 2 and 3, although actually indifferent, appears on the surface at least to be clearly differentiatable and determinable without indifference. The point is that the difference between finite numbers is the exception. In reality, in the world at large, we are constantly trying to determine differences between infinite cardinal multiples. In this instance then we would be faced with a bad infinity of indifferent choices either due to total indetermination or total determination. But the conceptless choice provides a very different view of indifferent differentiation. It determines that you can choose to such a degree that you can fill, almost, the chasm between a situation and a state, but that in doing so you must admit that your choice is an indifferent, i.e. conceptless, choice.[14] Even a choice can be indifferent, Badiou argues, as long as you accept, as Easton asks you to, the coherency of the deductive apparatus that allows you to choose, an apparatus by the way that Badiou has just proven to be indifferent.

Badiou's conclusion is that Easton's Theorem allows Being to be pronounceable only if it is 'unfaithful to itself' meaning not so much that conceptless choice brings a degree of determination to the very heart of indifferent, quantifiable determination, but rather that 'The un-measure of the state causes an errancy in quantity on the part of the very instance from which we expected – precisely – the guarantee and fixity of situations' (BE 280). This is significant not so much because Badiou uses conceptless choice to reintroduce the void into the state 'at the very jointure between itself (the capture of parts) and the situation', but because conceptless choice is the only solid 'proof' we have that singularity by virtue of the event 'exists'. Although intervention is a decision not a proof, if we decide to choose then mathematics can 'prove' the necessity of choice both in terms of what it can go on to do, and also in terms of how it can do all these things using the transmissible means of deductive reasoning.

In relation to this, Badiou speaks of a Cantor-Gödel-Cohen-Easton symptom (we didn't mention Gödel's role in this for the sake of brevity): 'That it is necessary to tolerate the almost complete arbitrariness of choice . . .' (BE 280).[15] By which he means that for there to be being as nature provable with only one decision, that the void exists, then there must be the event. Certainly such a decision requires a second decision, but this is a decision which is entirely justified retroactively in terms of the axioms of being, and totally demonstrable in terms of the norms of deduction available to classical logic. More than this, this second decision, that of conceptless choice, pertains to the first. The axiom of choice reveals, Badiou insists, the void at the heart of natural structure. This has the concomitant effect of being another demonstration of the justified nature of the decision, the void exists, and while on the one hand it introduces a permanent incoherence to the coherent structure of being, at the same time it also necessitates that choice be tamed by the deductive processes which determine being. Thus, two decisions exist, the decision of being, that there is the void, and the decision of the event, that you can choose randomly, or without concept, within the coherency of being. While even one decision presents a limitation to the imperialism of a constructible universe, these two decisions together have the benefit of relating to each other, even supporting each other, and of almost cancelling each other out. For a realist such as Badiou it is hard to imagine a more secure position than this. The thesis could be stated as follows: in the case of the norm, infinite cardinality, quantity is indeterminate. Yet these states can be known, deductively, as long as you accept that this knowledge is not sufficient in itself, not constructible. Instead, to prove the consistency of infinite cardinals, whose set of parts it is impossible to count using one-to-one correspondence remember, you need to accept the reality of the void and in so doing you cannot negate the implications of choice, which is an initially in-different event.

The conceptless choice of quantity means you can determine that there is a consistent set of parts for an infinite cardinal, even if, due to the nature of said cardinals, you can never count or measure the quantity of said set. Conceptless choice is not just how a purely quantitative relationality allows content neutrality to be the founding precondition of being and event, but more than this, quantity as such is conceptless. As Badiou notes: 'If the real is impossible, the real of being – Being – will be precisely what is detained by the enigma of an anonymity of quantity' (BE 282). Clearly, it is not just that relations of difference can be determined due to quantity alone, the founding theory of the quality indifference of being, but even when thinking of quantity, the abyss of choice means that quantity as such is indifferentiated. Not only can you, by virtue of set theory, say one element is different from another due to quantitative relation without recourse to quality or predicates, but you can then determine quantity as such without recourse to relationality as such, obviating any temptation to speak of the quantity of multiples as relational and so, potentially, possessive of some kind of quality.

Conditioning the indiscernible

Moving away from Easton and back to Cohen now, the first section of Meditation Thirty-three delineates how one can speak of the fundamental quasi-complete situation, *quasi*-complete due to Gödel's incompleteness theorems, in an entirely generic fashion so that it can be the discernible basis for the indiscernible element we are looking to condition. So if we start with the ontological condition of a situation as indeterminate, we discover that the 'intrasituational approximation of an indiscernible' (BE 358) needs to follow as closely as possible the resources available to ontology as such, namely set theory. To which Badiou adds, by specifying the absolute nature of the quasi-complete situation from which our indiscernible, generic and indifferent element will emerge as not constructed but conditioned. These are: to be an ordinal, the first limit ordinal, and the set of *finite* parts for α such that they are counted as one for *S*.

With this formal architecture in place, what is indiscernible can now be conditioned in such a way as part of it becomes generically discernible. This amounts to a more fully worked out sense of saying that our multiple is a multiple as such, or in general, and thus we can say of it certain things without ever saying what this multiple is. In effect, Badiou is taking Hegelian pure difference as such, adding a large amount of extra detail as to how two elements can be discerned, differentiated, in a quality neutral way, logically, and then saying such indifferent determination is good not bad. Badiou explains that a condition is a multiple, he calls it π, of the fundamental situation, called *S*, which is 'destined to possibly belong to the indiscernible ♀'. ♀ then is the material, while a condition is a mode of transmitting information about ♀. He asks the perfectly legitimate question at this point: 'How can a pure multiple serve as a support for information? A pure multiple "in itself" is a schema of presentation in general: it does not indicate anything apart from what belongs to it' (BE 362). The solution is that we won't work on the multiple as such, but on how information in general can be assigned to any multiple using the simple laws of differentiation. In particular, we can speak of π in terms of order and domination.

If we take π_2, for example, we can say that it is more restrictive than π_1 or more precise. It contains all of π_1 for example and gives all the information therefore on π_1 plus a little more, that pertaining only to π_2. This designates an order which is also an order of richness in terms of information. Each new number includes all the previous numbers plus a new number, if you will, so it has more information in it and is therefore more specific. For the rest of our consideration of conditions due to relational order we will use a bare schema of zeros and ones to be able to 'talk' about our indiscernible. So if we start with the string *<0,1,0>* we can say that the string *<0,1,0,0>* is richer. It gives all of the initial string plus an extra 0 at the fourth position. The second string is said to 'dominate' the first as it contains all of the first plus something extra. In this way we can 'make the nature of the indiscernible a little more precise' (BE 363). This is the basic definition of information and as one can see it is totally indifferent: it says nothing of quality or specificity.

Now we come to *compatibility*. To stop our strings of ones and zeros accumulating randomly in such a way as they themselves are indiscernible to our quasi-complete situation *S*, which would be useless as we need something discernible to be able to condition what is not discernible, these strings must obey the simple laws of *S*. So we cannot have two strings which contain a contradiction featuring together in the conditioning of ♀ as these are incompatible. Thus *<0,1>* and *<0,1,0>* are compatible because the second string contains all of the first. But *<0,1>* and *<0,0>* are not because the second does not begin by reproducing the first before adding extra information. We might note here that we said initially that our indiscernible multiple was indiscernible because it contained a contradiction which knowledge does not tolerate. Clearly then it matters when contradictions occur. When making an enquiry contradictions are disallowed. One does not connect to negative elements so one's string is made up entirely of positives. When that enquiry is related to knowledge, however, it can connect elements taken to be contradictory because their condition of relation is connection to an event, not connection due to compatibility. However, when one wants to fix the condition of ♀ one must submit to non-contradiction again as this is a law of *S*. This produces a clear sequence: noncontradiction due to contradiction ordered through noncontradiction.

So far we have dealt with order and compatibility, but for Badiou we must also have an element of choice. He says: 'a condition is useless if it already prescribes, itself, a stronger condition; in other words, if it does not tolerate any aleatory progress in the conditioning' (BE 364). This actually is the key to ♀ being generic. If we take the string *<0,1>*, then the string *<0,1,0>* reinforces *<0,1>* by saying the same thing and more. One could also speak of string *<0,1,1>*. It also says the same thing and more. But the 'more' of the two strings is incompatible, one insists that *0* is next the other that *1* is next. At this point, to proceed to the next largest string one has to choose which of the two incompatible next strings you want and this will then determine all strings to follow because they all have to reproduce entirely what came before. Badiou explains: 'the condition *<0,1>* admits two incompatible extensions. The progression of the conditioning of ♀, starting from the condition *<0,1>*, is not prescribed by this condition. It could be *<0,1,0>*, it could be *<0,1,1>*, but these choices designate different indiscernibles. The growing precision of the conditioning is made up of real choices; that is choices between incompatible conditions' (BE 364). What this sequencing does is allow for indiscernibility as such to be a specific indiscernible due to actual choice yet without ever stipulating what qualities one is choosing or not. All one asserts are that there will be choices and they will follow the rule of compatibility. There are, then, totally indifferent, choice-based specificities that can be extremely specific, but they will never become a constructible, named specific thing.

Conditioning is a manner of giving very detailed information about a multiple without ever saying what the multiple actually is. It resembles the as-if formula of Kant, if you like, but it is infinitely, literally, more specific than that, taking its inspiration more pointedly from Easton's Theorem. One can construct a whole infinite world of information about an indiscernible element both in general, *all* such elements will function in this way (materiality), and now specifically, *this* element functions in this way. All of this is down to four laws. The first is the law of materiality, what a set of conditions will be and how it will behave so that ©, the set of conditions, *S*, the quasi-complete situation, and π, the parts of the multiple ♀, all inter-relate in a stable fashion. The remaining three say, if we accept that an indiscernible can be conditioned due to discernibility, and thus in this fashion exists, this is what we can say about it. That it is ordered. That two conditions are compatible if they are dominated by the same third condition. That every condition is dominated by two conditions which are incompatible with each other. Thus we can say that every indifferent element is indifferently determined to the degree of its materiality, its order, its compatibility and our choice. This is a sentence that was impossible to conceive of before the publication of *Being and Event* and after it, nothing in philosophy can surely be the same again.

Indiscernible or generic subsets[16]

We have been able to determine the material for an indiscernible subset and information on that subset. Badiou presents this as π_1 belongs to ♀. This statement both says that the condition π_1 is presented by ♀ and what Badiou calls 'the same thing read differently' that ♀ is that to which π_1 belongs or can belong. We have now the minimal piece of atomic information, but naturally we are pushing for much more, a mode of collective conditioning. In pursuit of this, Badiou asks us to ignore ♀ for a moment and just concentrate on determining what a correct subset of any multiple is. To do this he presents two basic conditions. RD1 states that for a set of conditions to be correct, if a condition belongs to this set then all the conditions that the first condition dominates also belong. This also allows us to posit the void as the minimal condition of such a set. The second rule, RD2, encapsulates the laws of compatibility and domination. Given two conditions, there exists a further condition which dominates them, by which we mean includes all their information and adds information.

If this is how we define a correct subset, then can these two rules be applied to an indiscernible subset? Section 4, 'Indiscernible or Generic Subset', attempts to do just that by using a simple technique of classical logic, the excluded middle, to show what the indiscernible is by saying it is not the discernible. The basic logical operator is A ≠ ~A. If ~A exists, then ~ ~A means that A exists. Thus, if A exists, then ~A exists. Another way of looking at this is that Badiou is defining the rules of what is constructible so that something which breaks those rules exists within the constructed universe as the ~A or that which does not follow the rules. He is, in other words, turning constructivist thought against itself by forcing on it the possibility that something non-constructed is present therein definable systematically and procedurally by the laws of construction and how said element succumbs to those laws in the negative.

Following Badiou's reasoning, he says first take a set δ as discernible for *S* if there is an explicit property which names it completely. This is the formula λ(α). So that to belong to δ and to have the property expressed by λ(α) is the same thing. Not only do all the elements of δ possess this property, if any element possesses this property they belong to δ as well. Thus, if α does not possess this property then it does not belong to δ. This is how λ names the set δ by separating it. As we saw, this is the basis of a set as indifferent collection. If we extend this to two conditions of π_1, either π_2 or π_3 using the law of choice as we have just detailed it, we can say that only one element can belong to δ, we will say it is π_3. From this we can say that 'Since the property λ discerns δ, and π_2 does not belong to δ, it follows that π_2 *does not possess* the property expressed in λ', because remember to belong is the same as having the property, so if you do not belong you do not have the property. This is written as ~λ (π_2). The result is that 'if a correct part δ is discerned by a property λ, every element of δ (every $\pi \in \delta$) is dominated by a condition π_2 such that ~λ (π_2)' (BE 367). Thus, the indiscernible element of λ is part of the conditions of *S* because it dominates λ by not being compatible with it. This allows us to name a set in terms of domination, for example only containing 1s separates out the sets *<1>*, *<1,1>*, *<1,1,1>* as clear subsets. To negate this 'discerning property' we can propose the rule containing *0* at least once (BE 368). This dominates our first rule because when one adds a *0* then the first subset is over: *<1,1,1,0>* is the exterior of the subsets only containing *1*s because it contains a *0*. This means an exterior element can always dominate the interior of any subset. From this, Badiou makes the following deduction on the excluded middle albeit one much more sophisticated than simply saying if A exists then ~A is discernible as that which does not exist. He says:

> We can therefore specify the discernibility of a correct part by saying: if λ discerns the correct part δ (here λ is 'only having *1*'s'), then, for every element of

> δ (here for example, <*1,1,1*>), there exists in the exterior of δ – that is, amongst the elements that verify $\sim\lambda$ (here, $\sim\lambda$ is 'having at least one *0*') – a least one element (here, for example <*1,1,1,0*>) which dominates the chosen element of δ (BE 368–9).

For Badiou, the first implication of this line of reasoning is that we can define discernibility structurally, without any reference to language. Instead of speaking of properties, in other words, we can talk entirely in the abstract by saying that any condition outside the domination is dominated by at least one condition inside so that there is a simple, structural law of choice, either one or the other. The product of this mode of reasoning is: '*Every correct discernible set is therefore totally disjoint from at least one domination:* that is, from the domination constituted by the conditions which *do not possess* its discerning property' (BE 370). This has the concomitant result that if a set exists which intersects with every domination, here it intersects with both *1* and *0*, then this set by definition is indiscernible, 'otherwise it would *not* intersect the domination which corresponds to the negation of the discerning property' (BE 370). Working in this manner permits Badiou to construct at the very least a concept of indiscernibility or a means of saying what such a thing actually means within constructible language. For us it is the first formal description of what is meant by indifference within the world that is outside of ontology, a logical world of comparative relational differences where initially we thought indifference would be banned. This description of ♀ is as follows:

- ♀ must intersect or have at least one element in common with *every* domination relative to an inhabitant of *S* (in other words this is a constructible, intrinsic universe).
- There could exist other dominations, but they are not relevant to *S*, so indiscernibility is local not absolute.
- Indiscernibility is relative to *S* and so domination is also relative to *S*.
- 'The idea is that, *in S*, the correct part ♀, intersecting every domination, contains, for every property supposed to discern it, one condition (at least) which does not possess this property. It is thus the exemplary place of the vague, of the indeterminate, such as the latter is thinkable within *S*; because it subtracts itself, in at least one of its points, from discernment by any property whatsoever' (BE 370).

One important proviso is that this intersection is not empty. From this, Badiou is able to confidently say that he can speak to any inhabitant of *S* as regards the indiscernible element in terms of *S* and in a language that the inhabitant understands because it is the language of *S* as such. This is an essential rule for making the indiscernible element communicable (existent).

Badiou ends by first showing that this law is pertinent to any series whatsoever by giving various examples of it in play. The content of the series and the rule of its domination is totally irrelevant, it is indeed indifferent, all that matters is the law that: 'The generic set, obliged to intersect every domination defined by these properties, has to contain, for each property, at least one series which possesses it. One can grasp here quite easily the root of indeterminateness, the indiscernibility of ♀: it contains '"a little bit of everything", in the sense in which an immense number of properties are each supported by at least one term (condition) which belongs to ♀' (BE 371). The only limit on this law is consistency. So ♀ can't contain two conditions that two properties make incompatible, for example beginning with *1 and* beginning with *0*. Then '[f]inally, the indiscernible set *only* possesses the properties necessary to its pure existence as multiple in its material (here, the series of *0*'s and *1*'s). It does not possess any particular, separative property. It is an anonymous representative of the parts of the set of conditions' (BE 371).

To summarize, if we take the conditions of discernment, then we can construct out of them an indiscernible element. When it comes to the domination of a series, then every series admits to an element which is external to it and thus sets an exterior limit. Faced with a choice of π_2 or π_3 it chooses one or the other. As regards the indiscernible element, in terms of choice both dominations could be chosen. One cannot choose both in terms of discernment, but if choice exists then so does the possibility of choosing both. The choice is defined, it will happen, but what is chosen remains indeterminate. We have already seen that this occurs as regards the event because a subject chooses what is connected to the event, not what is compatible with a series. It is very important here that the indiscernible is not simply the incompatible, as the incompatible is just the domination of any series. Rather it is a choice that disregards incompatibility presented in the language of compatibility so at the very least we can say, within *S*, the indiscernible can be in some senses discerned by speaking only in the abstract and accepting the possibility of a choice wherein the specific nature of the choice is never known.

The existence of the indiscernible (Meditation Thirty-four)

Having presented a concept of the indiscernible, we now need to ask if there exist any examples of this concept. To exist means to belong to a situation, so do any indiscernible elements belong to a situation? Disappointingly, the initial answer is no. Badiou is unable to show any situation for ♀ to belong to *S* if π is generic through the law of intersection with every domination, without having to say that π intersects its own exterior. This is impossible so that whilst the laws of a situation do allow for the idea of an indiscernible, by the same gesture they also disallow its existence: 'since no generic subset exists in this situation, indiscernibility remains an empty concept: the indiscernible is *without being*. In reality, an inhabitant of *S* can only *believe* in the existence of an indiscernible – insofar as it exists, it is outside the world' (BE 373).

Yet all is not lost because while an indiscernible element does not exist for an inhabitant of *S*, as regards ontology in general this is not the case. For example, many elements in ontology are only observable from the outside, such as the cardinality of a situation which can only be observed outside the situation. More than this, ontology can define a set of the dominations of *S*, and go even further and do this in the abstract to such a degree that for their purposes 'there is no doubt that a generic part of *S exists*' (BE 374). I won't reproduce the proof here as that is not quite the point. Suffice it to say that all we are establishing is that ♀ is included in *S* but does not belong. As we have seen such excrescent elements exist ontologically speaking. Badiou restates the following law, that in ontology, '*there exists a subset of the situation which is indiscernible within that situation*. It is a law of being that in every denumerable situation the state counts as one a part indiscernible within that situation, yet whose concept is in our possession: that of a generic correct part' (BE 374–5). So it can be seen that many elements of a situation are observable outside of the situation, that is, exterior to the language comprehensible to the inhabitants of that situation. If we accept this, then it becomes relatively easy to prove the generic exists as a condition fulfilling RD1 and RD2, albeit not in a fashion that speaks to elements within the situation.

This then becomes the problem. Through ontology we know that ♀ exists, but we can never say to which *S* it exists. To make ♀ exist for a specific situation it must be added to an *S*. Yet to do this it must be appended to *S* using the language and resources of *S*, otherwise *S* will never know that ♀ has been added to it. This is because a fundamental law of any *S* is $\sim$ (♀ $\in$ *S*). To achieve this the very language of a situation needs to be altered so that you can name in *S*, 'the hypothetical elements of its extension by the indiscernible' (BE 376). This would allow an inhabitant of *S* to say if an indiscernible exists, these would be its properties, without saying it does exist, as for them it cannot. At the same time the ontologist can realize this hypothesis because they know that generic sets exist. Thus, the articles of faith for the inhabitant of a situation are real terms for the ontologist.

Badiou's excitement is tangible and infectious: 'The striking paradox of our undertaking is that we are going to try to *name* the very thing which is impossible to *discern*. We are searching for a language for the unnameable. It will have to name the latter without naming it, it will instruct its vague existence without specifying anything whatsoever within it' (BE 376). This is tantamount to saying we need a language of indifference, a seeming contradiction in that language is a differential mode due to specific quality. To build the impossible, a language of indifference *S* (♀), Badiou starts with the position that it must be constructed from the resources of *S* alone so it will be a constructible language of names. In that for *S*, ♀ inexists, the only elements of ♀ that exist simultaneously for the being of ♀, provable ontologically from the outside, and for *S*, are conditions. It was meeting the condition RD1 and RD2 that allowed Badiou to state '♀ is generic' (BE 374). And as we saw it was the same conditions that defined elements within *S* as well, in a general abstract form such that at least an inhabitant of *S* could conceive of ♀, even if the logic of their situation disallowed its existence. A name will bring together a multiple of *S* (which is a name) with a condition. This combination also allows us to think that every name is itself made up of couples of other names and conditions. This is the essence of every constructible universe, names plus conditions or words plus grammar if you prefer.

Badiou's formulation is that a name is a multiple whose elements are pairs of names and conditions, so if you want to define a new name you use an old name, combine it with a condition and hence form a new name. The problem here is of course circularity in that you presuppose that you already know what a name is, indeed that such a name exists as well. However, as Badiou points out, in ontology said circularity can be 'undone, and deployed as a hierarchy of stratification . . . it always stratifies successive constructions starting from the point of the void' (BE 376). This is why realism is essential as a supplement to constructivism when it comes to the language of indifference. The void negates the circularity of naming by having a lowest limit beyond which naming cannot trespass. There is, in ontology, a first name and that is the name of the void which is, by the way, indifferent. Thus wherever you start in the rank you are always basically dealing with the simple couple $<0,\pi>$, where *0* is the minimal condition of the name, and π an indeterminate condition. Without ever knowing what π is, or where it is in the scale, we can always say that if we need to we can determine its place indexed to the void, which conditions nothing, and π which conditions something. All that is required to make this operative is to presuppose that π is indexed to an element smaller than π, to which π functions as its domination. Badiou explains it best, as you might imagine:

> The definition ceases to be circular for the following reason: a name is always attached to a nominal rank named by an ordinal; let's say α. It is thus composed of pairs $<\mu,\pi>$, but where μ is of a nominal rank inferior to α and thus previously defined. We 'redescend' in this manner until we reach the names of the nominal rank *0*, which are themselves explicitly defined (a set of pairs of the type $<\emptyset,\pi>$). The names are deployed starting from the rank *0* via successive constructions which engage nothing apart from the material defined in the previous steps (BE 377).

Clearly we have to accept the realism of the void. It is essential to state that the void 'exists' to form a basis of removing the circularity from the assumed ostensive function of the first name or name of the name. Then we also need the language of construction for the void to speak to actual situations. Finally, we have to presuppose the foundation of all conditions due to logic.

The next question is, given this very strong determination of an indifferent language, differing from other theories of language because it presupposes the realism of the void and thus that truths can occur, is this determination intelligible and thus acceptable to actual situations? Does an indifferent language speak in such a manner that inhabitants of this world can hear it, even if they do not yet understand it? Having established that not all names in ontology exist in *S* and various other restrictions, Badiou then satisfies himself that a name for ontology (ordinals, pairs, sets of pairs and so on) is the same for *S*. From this he then proposes the project of constructing a situation to which the indiscernible will belong or working in such a fashion that the name creates the thing or *S*(♀) creates ♀ for *S*. Finally, through a set of procedures we will not repeat, he shows that it is possible to give names a referential value tied to the indiscernible using procedures and operations only acceptable to *S*, or to the constructible universe. To achieve this one basically follows the procedures for naming any element, the combination of name plus condition, but tied to the symbol ♀ which one supposes exists or accepts could exist or which one places in some form of conditional that *S* accepts and understands. This is of the order that if an indiscernible did exist, then it could be named using the referential pairing of name and condition as we would with any element that actually exists.

Again there is an apparent circularity here in that you determine the name μ by assuming you can tie it in to the name μ_1 even though μ_1 only exists as a referential determinant of μ. Again, this circularity is negated by unfolding the circle into the hierarchy of the names' nominal rank as we did before, counting all the way back to the void and determining the referential value of this actually valueless and inexistent name again. The formula Badiou gives here leaves us either with a generic pairing with the referential value {Ø}, which basically means there is a connection to the 'void' of the event, or simply Ø which means there is not. The entirety of this string of deductions is totally comprehensible to the inhabitant of S, as she is fully conversant with the 'if ... then' structure as determined by the conditions of naming for her constructible world. Thus, if an indiscernible did exist, we could name it and because we can name it in principle, in abstract and generically, in this sense it can be said to exist after a fashion. Which is what Badiou means by a generic extension: 'the generic extension by the indiscernible ♀ is obtained by taking the ♀ -referents of all the names which exist in *S*' (BE 380).

Extension

The simple question is, taking a situation *S*, where all names conform to the laws of discernibility and then classification, if we presuppose instead of a specific name, the general idea of naming as such, which is indiscernible in that it is indifferent, not because it simply does not 'exist' (it exists ontologically just not in *S*), can we extend all the properties of naming a specific object and say they can be extended to 'whatever' object? Are the conditions of naming, which is itself the condition of existence, generally applicable, abstractly viable, indifferent to any specific content? Can a constructible universe exist made up of elements which are indifferent as well as different, or is the indifferent language of which we spoke a language which no one understands and so is fundamentally non-communicable? Badiou does not want to change the language but he does want to retain the event and as we have repeatedly seen, the only way of registering the trace of an event is in *S*, in the constructible worlds of language. As Badiou says: 'The "nominalist" singularity of the generic extension lies in its elements being *solely* accessible via their names' (BE 381).

One of the issues raised here is the problem of invariance. Because ♀ is generic, one will never be able to determine it except by extending it in relation to *S* which is specific. Yet because one never knows the actual value of ♀, one needs something in *S* which never changes to be certain that in every case of ♀ it can be related through extension from *S*, or 'for every element α of *S*, there exists a name such that its referential value is α *whatever the generic part*' (BE 381). As it happens there is such a condition included in every grouping in *S*, the minimal condition of Ø. The named void as condition belongs to every situation but can never be specifically included as it is dominated 'by any condition whatsoever' (BE 381), a description very close to that of the indiscernible multiple, which is also determined by belonging, ontology, and can never be specifically included, because it can never be named specifically. The void conforms to both RD1 and RD2, and so is a correct part of every non-empty set in *S*, meaning it can function as the canonical name or if you like the generic name as such. For a third time we are confronted with a circularity in that the canonical name of α is defined on the canonical name of its elements, and again this circularity is negated by counting back to the minimum condition of the void. Badiou concludes: 'The canonical name of α is therefore the set of ordered pairs constituted by the canonical names of the elements of α and by the minimal condition Ø ... Now, and this is crux of the affair, the referential value of the canonical name $\mu(\alpha)$ is α itself *whatever the supposed generic part*' (BE 382).

This position solves Badiou's original problem as to how one can make sure that for every *S*(♀) all of *S* is included in ♀, because that is what extension means, by virtue of the canonical name which belongs to every situation irrespective of the generic nature of that part. Said stipulation solves the nominalist problem in *S*, in that every name in *S* is reducible to a name whose extension is the pure multiple itself. Which is tantamount to saying that at the base of all names is a real, which is the void. This real is itself indifferent, but because it is real it is also invariant to all names. As it is invariant to all names, if it can be shown that any generic name as such must accept that the void belongs to it, if it is a non-empty set, then it can be proven that any generic name is an extension of *S*, it talks using the same process of naming as *S* because, as we saw, the idea and existence of the indiscernible was indexed to the foundational role of the void set. Now we can say that the idea of the indiscernible, the existence of the indiscernible, and the language of the indiscernible all 'exist' in that they are tied to or indexed by the indifference of the void set. Thus, a language of indifference that is ontological, empirical and constructible is eminently possible, but for now suffice it to say that we can at least admit a limited vocabulary of indifference that can be comprehended in the constructible language of difference.

Is there a name for the discernible such that it can be said to exist?

The next question is does an indiscernible that is an extension of *S* actually exist? Badiou has already shown that indiscernibles are not just possible but, in ontology, actual, but as regards *S* the only things which exist are names. We have said that naming ♀ in *S* is possible in general, or generically, but is there a single instance of this: an actual name for ♀? In principle this is impossible. Naming is, by definition, discernment and so an actual name for ♀ means making something indiscernible discernible. We have already stated that *S* can discern indiscernibles of this order all day long and indeed loves to do so. The quest now is for a fixed name for ♀ that however is not discernible. Does such a name exist? Well, the only reason to discern a name in knowledge is to classify it, so if we can determine a name that cannot be classified we can then say that said 'name' is indiscernible but existent. This is, in fact, Badiou's definition of the truth of an event. So that while Badiou gives us a technical proof of this, we can satisfy ourselves with the fact that yes, such a name is possible. As he concludes after his formal calculations: 'We thus find ourselves in possession of a name for the indiscernible, a name, however, which does not discern it! For this nomination is performed by an identical name whatever the indiscernible. It is the name of *indiscernibility*, not the discernment of an indiscernible. The fundamental point is that, having a fixed name, the generic part *always* belongs to the extension' (BE 383).

Basically, the name of the indiscernible is a kind of place-holder such that, whatever the specific situation, there is a means of naming the indiscernible element within said situation as part of the situation to such a degree that it is also an element, it exists. As he concludes: 'This is the crucial result that we were looking for: the indiscernible belongs to an extension obtained on the basis of itself. The new situation *S*(♀) is such that, on the one hand, *S* is one of its parts, and on the other hand, ♀ is one of its elements. We have, through the mediation of the names, effectively *added an indiscernible to the situation in which it is indiscernible*' (BE 383–4). We can see that for ♀ to exist as such it must be an element, so it must belong, and it must be a name, so it must also be included. The only way an indiscernible element can do this is if it belongs to itself through the process of generic naming. This is how the self-predication of the event in ontology, which is totally banned, can, through the process of naming, exist as forbidden in the world of *S*. All of this is indexed to the indifferent name of the void. This name is a name, but it names nothing. As a name, it belongs to every situation and so is an element. Yet as an element it does not succeed. The axiom of foundation bans, so we thought, self-predication, but because foundation is tied intrinsically to choice – it is indeed foundation that means we can have the indifference of random choice – ironically it is foundation that allows for self-belonging elements to exist as long as they exist in general. Another way of looking at an event then is a licence for self-predicating elements to exist, if and only if they accept the generic name of indiscernible. So, the indiscernible is indexed to the event, and the event is under licence from the indiscernible, indifferent name.

Having proven the first two points, the third, which historically has been the most tricky, do indiscernible or singular elements remain thus when named and included, is actually rather easy to resolve. If we show that ♀ exists and does so in a way that is understandable to an inhabitant of *S*(♀) this still does not mean that ♀ becomes discernible. If, as Badiou says, 'we are looking for a concept of *intrinsic* indiscernibility; that is, a multiple which is effectively presented in a situation, but radically subtracted from the language of that situation' (BE 386), then ♀ is precisely that for the reasons given. Indeed, Badiou invites us by now not to worry about the proof he gives which he says is simply 'indicative'. So we will not. Suffice it to say that the multiple ♀ is indiscernible for an inhabitant of *S*(♀) as 'no explicit formula of the language separates it' and if it cannot be separated, then of course if cannot be differentiated, ergo it is indifferent. So that for an inhabitant of *S*(♀) 'there does not exist any intelligible formula in her universe which can be used to discern ♀' (BE 387). I cannot overstate the significance of this apparently technical formulation for our wider consideration of discourse through what we are calling communicability. In that Foucault's basic condition for any statement in discourse is intelligibility, what Badiou shows is that within discourse it is possible that a statement exists which is not intelligible, yet which belongs to discourse. This is a proposition with immense implications which we cannot pursue here. Instead, returning to Badiou's 'more modest' proposal, we hear him declare with justifiable glee: 'We have obtained an *in-situation* or existent indiscernible. In *S*(♀), there is at least one multiple which has a being but no name. The result is decisive: ontology recognizes the existence of *in-situation* indiscernibles' (BE 387). Here in-situation should first be understood as in a situation, but it should also be taken in the same manner is in-difference, in that indiscernibles are present to the situation by not being named specifically within. In that they remain indifferently named their location is in-situation by the same measure. Badiou ends with a summation of this remarkable idea of the generic name of the indiscernible multiple in the language of a constructible universe *S*.

> The indiscernible subtracts itself from any explicit nomination in the very situation whose operator it nevertheless is . . . What must be recognised therein, when it inexists in the first situation under the supernumerary sign ♀, is nothing less than the purely formal mark of the event whose being is without being; and when its existence is indiscerned in the second situation, nothing less than the blind recognition, by ontology, of a possible being of truth (BE 387).

The generic, indiscernible or indifferent inexists in ontology so as to be named ♀, allowing it to be discerned but without the need for classification. Then it exists in language but remains indiscernible. The result is a two-part demonstration of the conception of being and the proof of its existence. As regards ontology, the indiscernible marks inexistence, gives material presence to the concept of an event, even if it cannot exist in ontology. This inexistence is still something, a presence, a potential and so on. As regards logical constructivism, the indiscernible instead gives existence to the event even if it cannot mark it.

In closing on the generic, we can say with great certainty that Badiou has made a major entry into the history of thought due to two basic elements of set theory. The first is foundation due to the void of being, the second is infinity due to the inconsistency of infinite multiples of infinite beings (they do not need to form a One). We have an ontology which does not depend on totality to be totally stable (actual infinity), and which has a foundational minimum. The surprising thing is what Badiou does with this discovery. A sensible response would be to show that total stability exists and philosophy, as the science of being, is back in business so to speak. Instead, Badiou uses stability of being and beings in the world as the discernible and constructible backdrop to reintroducing the validity of the event each as concept, as real, and as nameable. To achieve this, he has to remove set theory from Gödel via the reintroduction of the axioms of foundation and choice. He has to insist on the realist position of at least one primitive, the void, and by virtue of that void also the actually infinite (our second existential seal). And then he has to use Cohen's proof of the generic to show that the event, which cannot exist ontologically but can be named, can then trade on that name to exist in the world as that which is named as the indistinct, the indiscernible, the generic or the indifferent.

While this gesture is decisionist, one must decide to believe in the event and choose it through naming it, and mired still in the main project of a philosophy of difference since Hegel, that radical difference as such exists to a degree that stability can be broken, it does not suffer from the aporias of that philosophy because, due to set theory, it has a provable axiom of minimum and of actual infinity. Something his predecessors never had. That said, none of this would be possible without indifference which accompanies Badiou's thinking every step of the way from the primal in-difference of the void of being, through the indifference of multiples, right up to the extension of indifference into the heart of differential language in the form of the generic name of the indiscernible. Yet this entire achievement, and it is immense, even the demonstration of the event, is all in the service of one concept we have barely even touched upon. I am speaking, of course, of the subject. With one chapter to go, the subject makes a dramatic, last minute intervention!

8

Forcing: Truth and Subject

Only a truth is, as such, indifferent to differences.[1]

Theory of the subject (Meditation Thirty-five)

Events cannot be seen to occur in worlds without the intervention and fidelity of agencies willing to participate in the point-by-point process of enquiring after multiples to see if they can be seen to belong to the indiscernible multiple that the event makes possible on which they decide yes or no. This is Badiou's thesis anyway. These agencies Badiou chooses to call subjects, developing from his first masterpiece *Theory of the Subject* (1982) whose title he appropriates for this meditation only so as to criticize his first early attempts therein.[2] If you compare what he says about subject in1982 with what he says six years later, in reality the two positions bear little in common. (I will speak more of *Theory of the Subject* in our second volume as the early work has more in common with *Logics of Worlds* than it does with *Being and Event*.[3]) The new Theory of the Subject goes as follows: 'I term *subject* any local configuration of a generic procedure from which a truth is supported' (BE 391). This is the opening salvo of the thirty-fifth meditation. We can satisfy ourselves that we have allowed the luxury of fully-defining what a generic procedure is. Badiou commences to define the legitimacy of his resurrection of subjectivity by carefully stipulating all the things the subject is not, or all the myriad theories of the subject that he regrets. Thus, we learn a subject is not a substance. It is not a void point. It is not transcendent. It is not invariable to a presentation (the subject is rare). All subjects are qualified by the conditions they are subjects of – love, politics, poetry and science. Finally, a subject is not a result or an origin vis-à-vis the event. Rather 'It is the *local* status of a procedure, a configuration in excess of the situation' (BE 392). These then are interdictions on the subject as regards philosophical and ontological presuppositions we may have applied given freedom to do so.

Badiou moves to stipulate exactly what subjectivization is but perhaps we should commence with the summary of this list:

> Subjectivization, aporetic knot of a name in excess and an un-known operation, is what *traces*, in the situation, the becoming multiple of the true, starting from the non-existent point in which the event convokes the void and interposes itself between the void and itself (BE 394).

Dense though this definition is, it does allow us to present a fairly straightforward typology of what Badiou takes to be the process of subjectivization.

- A subject is local.
- It is tied to the configuration of a generic procedure.
- Its habitat is truth not knowledge.
- The configuration of a generic procedure is in excess of the situation.
- Said excess is due to the problem of the generic name for an event in the constructed language of the situation.
- This is the result of operations that are un-known because they concern, for example, the immense presence of indifferent and thus generic elements in ontology, which cannot present themselves within the constructible sets of language.
- Subjectivization is a process, it traces, its first job is discernment, even if it is the discernment of the indiscernible. A subject, in short, follows the marks of an element that can be 'discerned' without being classified. This is basically the definition of a generic name.
- The process that the subject traces is the means by which the event, which is a self-belonging and thus inexistent non-being closed element, starts to become discernible as a multiple, and eventually begins to generate infinite 'new' multiples.
- The original point of the event, and the event is punctual, is inexistent because it has no means of presenting itself to being as such or being an existent being in the world.
- The starting point for all of this, although it is wrong to think of it in any form of causal sequence, is the means by which the event brings together the void and the world.
- It does so due to its double status as an ultra-one, opening a space between the void and itself, which is the first gesture of its constructability: it can become a collection by placing itself as a part of itself conjuring a singleton not out of nothing but between nothing and its own inexistence.

The last point is the most complex and difficult to accept, but we are confident now that we have done all one can to show the legitimacy of this claim, *the* claim of the whole event-subject conglomeration. So, the foundations for the possibility of there being a subject are sound due to the incredible innovation of Badiou's suturing of philosophy to Cohen's theory of the generic and its concomitant assertion of forcing.

If we return to the detail of the first section on subjectivization we can now pass over much of this as taken as read. Badiou starts by emphasizing the double nature of the subject (as we saw the subject must be so due to the ultra-one status of the self-belonging event which is always Two). The double nature of the subject is the name of the event due to an intervention on an evental site (nomination) and 'the operator of faithful connection' which pans out across the process of enquiry echoing the process of knowledge construction through conditions. Specifically, subjectivization is 'the emergence of an operator' meaning one could retain all of *Being and Event* without accepting the subject. The operator clearly emerges after an intervention and as we saw occurs as a Two. It is directed both to the evental site and to the evaluative processes of enquiries due to the generic procedure. Subjectivization always intervenes on the event from the position of a situation. It is a constructible not strictly ontological concept. He calls this a 'special count' different from the presentative and representative counts we have thus far considered. A subject only counts what is connected to the name of the event. We might add that the subject is the only agency that can count these names as they have no being and they do not exist, they are not presented and they are not represented. This is a strong position. If we accept both that the idea of the event is consistent in ontology and the existence of the event possible in language and that in both instances the event cannot be named as such, then some agency must be able to name the event. So perhaps the problem here is less to do with the existence of such an agency as naming said agency a 'subject' when indeed it could be called any number of other designations. The choice of the term subject, therefore, is clearly strategic on Badiou's part as regards enhancing the evental and then communicable status of his idea, by which we mean at the time such subjects were basically impossible but that over time the term subject has been accepted because it allows the philosophical community of Marxist and Lacanian inflected thinkers in particular, to say and do things they want to say and do.

Now we turn to the Two of the subjects who have a name. In each case we are within a named situation, an overarching discursive conglomeration. So we have the church, the party, ontology, music and so on. These discursive constructions subsume, all the time, what they take to be new names but in fact they are mere modifications. Thus, when a subject to truth occurs, Paul for the church, Lenin for the party, Cantor, Schoenberg and so on, it 'subsumes the Two that it is under a proper name's absence of signification' (BE 393). What they initiate in the 'one' of such discursive consistencies is a split of the Two between the name of an event and the beginning of a generic procedure. So we have death of God and the Christian Church, revolution and Bolshevism, infinite multiples and set theory, the tonal system and serialism. In each case, we have a proper name, a person, but what this name designates is not that they intervened or that they were faithful, this is how most might take this view on the subject but that would be incorrect. Instead, the proper name, Cantor for example, simply oversees the 'advent of their Two' or the articulation of intervention with fidelity. Badiou constantly warns in his typology of events, subjects and so on how unless intervention happens in this way, events will not take place, truths will be lost, subjects will not be subjects and so on.

The subject then is named as the one who oversees 'the incorporation of the event into the situation in the mode of a generic procedure' (BE 393). That this results in an absolute singularity, real change, an actual event, is designated by the in-significance of the proper name in relation to language as such. This specific but as yet in-significant proper name is generic in the situation but it is singular in terms of material presence. At the same time, it is directly connected to the void as such so that 'Subjectivization is the proper name *in the situation* of this general proper name. It is an occurrence of the void' (BE 393). Such a statement is totally meaningless outside of the logic of indifference. In the situation, the indiscernible is simply indiscernible. Only if logic and ontology are articulated through the Two can one differentiate between indifference in general in ontology, generic indifference as such, and a specific indifference in an actual situation. Finally, it is undeniable that the subject serves truth. Badiou says explicitly that generic procedures assemble truth, that subjectivization is the making possible of a truth. It does so because it allows 'the evental ultra-one to be placed according to the indiscernible multiplicity . . . that a truth is' (BE 393). The resultant proper name bears the traces of this process being the trajectory of truth. In other words, the point of the subject is truth.

Chance

Having established what subjectivization is, its relation to the Two, and its service of the truth, we move on to the role of chance in the event. For a study concerned with the event, there has been little up to this point on chance aside from Badiou's reading of Mallarmé.[4] Finally, however, he concedes that all generic procedures, although systematically conditioned, depend on chance encounters. If fidelity is a series of enquiries after the positive or negative connections of generic names to events, an enquiry cannot encounter an event, it can only test the evental status of an encounter and this chance encounter must, according to Badiou, occur due to subjects. Thus, if subjects articulate intervention and generic procedures, they do so by being available to encounter things entirely by chance. Hence we can assert that: 'The procedure is thus ruled in its effects, but entirely aleatory in its trajectory' (BE 394). Chance is an essential part of the procedure, this much is clear, but 'This chance *is not legible in the result of the procedure*, which is a truth, because a truth is the ideal assemblage of "all" the evaluations, it is a *complete* part of the situation. But the subject does not coincide with this result' (BE 394). On the local level, the subject is involved in nothing but illegal encounters. This means that as regards how these encounters are rated, positively or negatively connected, the subject is 'qualifiable, despite being singular: it can be resolved into a name (e_x) and an operator (□)' (BE 395). Yet at the same time, as regards its multiple-being, all the terms which can be indexed by any enquiry, it remains unqualifiable tied, as it is, to pure chance. Cantor can be a totally qualifiable name, originator of set theory due to the discovery of actual infinity, and yet remain the unqualified subject of that event.

Badiou explains this paradox as follows: 'Of course, an enquiry is a possible object of knowledge. But the *realization* of the enquiry, the enquiring of the enquiry, is not such, since it is completely down to chance . . .' (BE 395). So that whilst knowledge can enumerate elements of enquiries afterwards and even index them to a name, Cantor, Schoenberg, Badiou, it can't anticipate in the moment the possibility of the grouping of these singularities as they are encountered. In a wonderfully damning phase Badiou concludes: 'Knowledge . . . never encounters anything. It presupposes presentation, and represents it in language via discernment and judgement' (BE 395). After two decades working in the Higher Education system I have no choice but to concur. In contrast to knowledge being all form due to presupposition of matter, the subject is all matter (terms of enquiry) without any form prescribing said matter.

Faith

Inasmuch as the one-truth is indiscernible to the language of the situation, because the subject is a local configuration of the procedure, we can also say said truth is equally indiscernible for them. Truth is global while the subject is local, so there is a profound incommensurability between the two. This disjunction is important to bear in mind to avoid characterizing the subject in terms which do not belong to it. A subject can only encounter, for example, what is presented in a situation. Yet we know that a truth is un-presented in a situation. So a subject never encounters truth or truths as such. A final restriction is that the subject can't make a language out of anything other than combinations of the supernumerary name and the already existent language. Badiou sagely concedes that we have no guarantee that this odd bricolage is sufficient to in any way capture a truth. 'It is absolutely necessary,' he says, 'to abandon any definition of the subject which supposes that it knows the truth ...' (BE 396). This stark reminder tells us that truth is always 'transcendent to the subject' (BE 397), so that the only agency the subject has as regards truth is the faith that the event occurred and their confidence that things will now be different.

If a subject is gifted with nothing more than a 'knowing faith' called 'confidence', then the nature of this confidence is important to say the least. There is an almost Heideggerian slant here of employing folk-language to access complex ontological truths but there is nothing Heideggerian about the systematic definition of what confidence is. Central to confidence is faith in the Derridean 'to come' and indeed the future anterior is by far the defining mood of all truth procedures from the axiomatic method, through the relation between intervention and the event, right up to confidence in the truth to come.[5] The operator of fidelity, which is not the subject remember, works to discern connections and disconnections between multiples and the name of the event. This results only in an "approximation of truth" whose effects are to come in truth. The belief in the to-come of a truth operates under fidelity to the name of the event: the revolution has arrived, now we must make things different. It does this by combining the language of the situation with the fidelity to the event which is not of the situation. Thus a 'finite enquiry therefore detains, in a manner both effective and fragmentary, the being-in-situation of the situation itself' (BE 397). We have come across this formulation twice before when we spoke of the presentation of presentation as such and also the generic ♀. In each case the logic is the same.

There is a certain self-reflexive self-belonging to every element of ontology due to indifference. Indifference allows for one to speak in general about ♀ as such from the perspective of ♀. This is naturally because indifference is operationally akin to difference, yet stripped of specificity. In this fashion, although we have studiously avoided such terms, indifference is in-between or liminal to some degree. Here Badiou falls back on the median status of indifference due to the Two of the ultra-one and of course the desperate desire to be able to translate the event into effects discernible in the situation even if they are discerned as the generic possibility of in-discerned elements. So he says of the to-come there is the discernible nature of these enquiries in relation to knowledge, built as they are out of the same generic conditions as knowledge (name and condition) in accordance with RD1 and RD2. Yet at the same time 'it is a fragment of an indiscernible trajectory' (BE 397), so an aspect of the to-come will always remain unknown. These technical details are the foundation for what Badiou terms faith: 'that the operator of faithful connection does not gather together the chance of the encounters in vain. As a promise wagered by the evental ultra-one, belief represents the genericity of the true as detained in the local finitude of the stages of its journey' (BE 397). It is in this way, in succumbing to this belief, that a subject to truth is born.

Names

A subject manifests her faith primarily in the generation of names. These can be neologisms or 'old' names loosened from their moored articulations. What is significant about these names is that they do not have a referent in the situation even if they already exist in the situation doing another job. Instead, the referent of these names is 'to come' when presented in the new situation. 'Such names "will have been" assigned a referent, or a signification, when the situation will have appeared in which the indiscernible . . . is finally presented as a truth of the first situation' (BE 398). Names in waiting if you will. The names are a kind of shibboleth, Badiou explains, in that for those who have no faith in the new situation to come, they are empty and meaningless. They are the new language, therefore, of a new subjective community. There are many problems here as regards these names which I think Badiou is aware of in his later work, so rather than debate these problems we will focus solely on the suspended nature of these names as that remains pertinent.

Badiou says: 'The names generated . . . by a subject are suspended, with respect to their signification, from the "to-come" of truth' (BE 398). Suspension results in a private language internal to a situation subject to a condition but of an as-yet incomplete generic part. The names must be of this order because it is impossible to anticipate or represent a truth. A truth is not manifest in its names, but in the encounters of the enquiry with respect to the situation in question. Truth is a meeting, a process, an activity; it is not for capture by nomination as such. 'Consequently, the reference of the names, from the standpoint of the subject, remains for ever suspended from the unfinishable condition of a truth' (BE 399).

The combination of the generic and suspended reference founds an idea of a form of indifferent naming which Badiou explains in the following process of formulation: 'All that one can do is say if a term, when it will have been encountered, turns out to be positively connected to the event, then this name will probably have a referent, because the generic part, which is indiscernible, will have this or that configuration or property'. This is clearly indifferent naming because it speaks of referential nomination in a quality-free, because content-less, situation. Badiou goes on to say of such a language: 'Language is the very being of truth via the combination of current finite enquiries and the future anterior of a generic infinity' (BE 399).

Again another useful summary; in effect the whole meditation is a kind of summary. So, here are the qualities of Badiou's second foray into a theory of the subject.

- A subject is concerned with the generic indiscernibility of a truth.
- It accomplishes this truth, Badiou's term, in discernible finitude.
- This is achieved by nomination using a suspended referent, suspended in terms of its specific mode of reference, but also in the sense of suspended 'from the future anterior of a condition' (BE 399).
- This makes the subject both the real of a situation, due to the enquiring of the enquiries (enquiry as such is real in that is apodictically presupposed by all constructible systems), and a hypothesis that said enquiries will 'introduce some newness into presentation' (BE 399).
- 'A subject emptily names the universe to-come which is obtained by the supplementation of the situation with an indiscernible truth. At the same time, the subject is the finite real, the local stage, of this supplementation' (BE 399–400).
- So when we say their nomination is empty, what we mean is it is full 'of what is sketched out by its own possibility' (BE 400). So not truly empty just full of indifferent, because as-yet content-less, names.
- Badiou summarizes all he has to say of the subject so far as follows: 'A subject is the self-mentioning of an empty language' (BE 400).

The reader would be correct in their observation that this makes a subject a self-predicating entity strictly forbidden in ontology, hence subjects exist in worlds only, *and* in constructible universes, meaning something of the subject will always pertain to the real of the void due to the possibility of choice, taken in terms of the axiomatization of choice by set theory.

Forcing (Meditation Thirty-six)

Badiou asks, at this very late hour, if the subject-language is 'separated from the real universe by unlimited chance' (BE 400), how can it ever be possible for one to declare said language to be veridical in terms of knowledge? For the external witness on the subject-language all the statements made by said subject are meaningless. All the subject has in response is their faith in the truth to come of this self-mentioning language empty of reference. Leading to the very legitimate question that is it not the case that the infinite incompletion of a truth denies any chance of evaluating said truth within the situation? Bringing us back to the basic problem of the whole study. Even if we can say, after Being, that events 'exist' to such a degree that they can generate subjects out of their belief in truth, can they ever talk to the rest of us in such a way as this truth affects the current state of presentation? This, after all, is the totality of Badiou's aim: to change the current state of things. If truths due to events do not change the situation we are in, Badiou has no use for them. He therefore needs one last formulation in the dying light of a very long day so that the entire study has not been worthless. That concept is forcing and it is due to what Badiou calls a fundamental law of the subject.[6] To have a statement of subject-language 'such that it will have been veridical for a situation in which a truth has occurred' (BE 400), the whole point of the book as I said, is because a term of situation exists which both belongs to that truth and maintains a relation with the 'names at stake in the statement'.

The relation in question is determined by the encyclopaedia. Meaning that the law 'amounts to saying that one can *know*, in a situation in which a post-evental truth is being deployed, whether a statement of the subject-language has a chance of being veridical in the situation which adds to the initial situation a truth of the latter' (BE 401). In this regard, it is enough that one such term exists that is both linked to the statement by a relation that is indiscernible in the situation. If you can find such a term, then the fact that it belongs to truth yet is legible in the situation means it will 'impose the veracity of the initial statement within the *new* situation' (BE 401). The result of imposition is not only that said statement is now legible to a situation, but that it is legible within a new situation! This is precisely the combination Badiou has been looking for and which he finds again in the work of Cohen under the heading of 'forcing'.

'I will term *forcing* the relation implied in the fundamental law of the subject' (BE 403), such that:

- A term of the situation forces a statement of the subject-language so that the veracity of this statement in the situation to come (the new situation) is equivalent to the belonging of this term to the indiscernible part that results from the generic procedure.
- This being the case said forced term belongs to the truth.
- By which we mean during the subject's 'aleatory trajectory' this term was encountered and tested positive as regards connection to an event.
- Said term forces a statement if its positive connection forces the statement to be veridical in the new situation. Which is another way of saying events can occur which never become veridical, and not all singularities encountered will test positive so that they can become veridical.
- Importantly, forcing is a relation '*verifiable by knowledge*' in that it combines a term of the situation and 'a statement of the subject-language (whose names are "cobbled-together" from multiples of the situation)' (BE 403).
- By the same gesture what is not verifiable by knowledge is if the term that is forced belongs to the indiscernible. That can never be known, it can only be encountered. In other words, you cannot 'force' forcing (cannot pre-prepare for an event) you can only force after an event has been accidentally met and systematically tested.

As we can see from our very first condition in this list, forcing occurs subsequent to and by virtue of the generic. So that while forcing produces a term which cannot be indifferent, it can be known,for example, that it can be forced is due to its being the result of indifferent generic procedures. Whether forcing produces a subject and changes the world we will put on hold for now, but what we can say is that forcing is the agency by which indifference is not merely extended into the world of communicability where it is effectively banned, but that as regards forcing, an act of violence against the indifferent term, for that is what an indiscernible term is, this means that an indifferent element can become discernible. From Agamben one learns that all differential economies, and Badiou's is certainly one of those albeit the most complete and sophisticated we have yet encountered, can be suspended due to indifference. Badiou is arguing that the obverse is true: an indifferent element can be forced to be differentiated. Yet why would one want to do this? Indeed, why does one need to when the concept of indifference as such, as a bad thing which difference can cure, is the central tenet of the philosophy of difference, since Hegel, Badiou is so effectively overcoming? The particular point here is that a forced indifferential element, when it becomes discernible, is without classification. So that while classification depends on discernment, discernment does not guarantee classification. A forced, to-be-differentiated, indifferent element does nothing to discernment except disallow discernment from becoming a class. This is perhaps the supreme irony of Badiou's application and interpretation of set theory. His use of set theory in ontology precisely allows for an indifferent multiple to enter into a constructible language as a multiple known because it shares no properties in common with any other already presented elements – most people's sense of what a set is.

All of this leads to the neat, impredicative formula: 'A subject is a local evaluator of self-mentioning statements', who knows that statements are either certainly wrong or possibly veridical in relation to the will-have-been of *one* positive enquiry. One can say of this subject that they are at the intersection of knowledge and truth due to their personal private language. It is a language so it can be known, but it is self-mentioned and of suspended reference, so it is currently indiscernible. Any subject can force the veracity of a statement of its language for a situation to-come. So that 'A subject is a knowledge suspended by a truth whose finite moment it is' (BE 406). While a subject cannot know truth, they can know if a truth can become, in the future, a veridical statement. Badiou summarizes the subject after forcing in the following fashion:

> Grasped in its being, the subject is solely the finitude of the generic procedure, the local effects of an evental fidelity. What it 'produces' is the truth itself ... It is abusive to say that truth is a subject production. A subject is much rather *taken up* in fidelity to the event, and *suspended* from truth, from which it is forever separated by chance (BE 406).

A subject puts an end to the generic indifference by forcing a statement, which is generic, to be promised to the veridical through a future language. That said, the subject through forcing can 'authorize partial descriptions of the universe to-come' because they know, at the very least, which statements that they have encountered have a chance of being veridical. This is still rather generic but one can see in it the process of becoming differentiated at the generic level. There is decision of difference here, these statements can become veridical these others cannot, but there remains a level of indifference even at the violet hour because the subject does not know which of the potentially veridical statements will come to be veridical. We are left with the concession: 'A subject measures the *newness* of the situation to-come ...' (BE 406). All of which under contract due to the following three provisos.

Undecidability: As the following meditation concerns undecidability we will simply say here that while a truth is subtracted from knowledge it does not contradict it. If a statement is undecidable to the encyclopaedia of the situation, it will never be known and so will never be veridical. Thus, while for Badiou it is acceptable that a statement can become veridical or that it is proven to be erroneous, that a statement remains undecidable halts the process of subjectivity effectively. If the subject is suspended from the truth to come, this is only if the truth to come is not in suspense of undecidability but is eventually decided upon.

This process gives the subject an additional role which is actually a crucial calling for Badiou's sustained commitment to classical logic, decisionism, dialectic and the law of the excluded middle. For him, as regards the truth of an event through the fidelity of a subject, contra-Derrida, any degree of undecidability is intolerable up to the point that he now defines the subject in terms of its ability to decide: 'that which decides an undecidable from the standpoint of the indiscernible. Or, that which forces a veracity …' (BE 407). How does a subject do this, as in fact it sounds impossible? Badiou lays it out in condensed form so we will break it down into clear steps:

- From within the being of the subject, an enquiry is reported where a term forces a statement that is connected to the name of the event.
- Said term belongs to an indiscernible (event).
- As a subject forces a statement we will know, at some point, if the statement is veridical or erroneous in the situation that results from the addition of the indiscernible.
- In that situation to come, which is truth, 'the undecidable statement will have been decided' (BE 407).
- Even Badiou contends that this simple yes or no to come is 'quite remarkable' in that it 'crystallizes the aleatoric historicity of truth' which would probably be presented by most thinkers as chaotic, indeterminate, impossible to know and so on, into a simple yes/no answer.[7]
- That such a reduction to Two, yes/no, can occur is due to the nature of enquiry and connection: 'It *happens* that such an undecidable statement is decided in such or such a sense' (BE 407). The Two presents us with a radical reduction of indifference as well. While we speak of an indifferent statement now, it will become a differentiated yes or no statement eventually. Indifference, on this reading, is still a facilitating mechanism for Badiou, as it was for Hegel and Deleuze before him. Indifference facilitates the impact of

> an event on our world in the form of a truth that changes what belongs to our current situation. That is all it is. It will die at that point, suspended literally by its neck from the rope, not of a question mark, as Lowell famously had it, but of an answer.

As the next proviso considers Badiou's failure in terms of *Theory of the Subject* as regards what happens to the terms of a situation after a new situation has come about, and the relish he had then for speaking of their destruction, a position he no longer agrees with,[8] we need not spend any more time on it knowing the completion of the subject is itself 'to come' in Badiou's work.[9] This leaves us with Badiou's closing comments on the third proviso he designates 'c', inexistence, which we can also pass over as they pertain to what happens to inexistent elements in relation to a new situation and how inexistence is an element of the ancient situation, something that need not concern us here. So we can move rapidly to Badiou's overall summary of the subject as that which, as a finite instance of truth, performs three operations.

1. It is the discerned realization of an indiscernible.
2. It forces decision and disqualifies the unequal.
3. And finally it saves the singular.

Indiscernibility, decision and singularity compose what Badiou calls a subject: 'By these three operations, whose rarity alone obsesses us, the event comes into being . . .' (BE 409).

The proof of forcing

The penultimate meditation describes in great technical detail the 'proof' of forcing within set theory. Even so, Badiou concedes this is a mere overview of Cohen's strategy which is too complex to reproduce in any detail. In one sense, we can state that just as there is a strong proof for the generic, so there is a strong proof for forcing and leave it at that. It is now clear what forcing is, a statement is forced onto the situation by virtue of a generic procedure which has been extended into the situation but which is fundamentally indiscernible therein. Forcing basically says, if this indiscernible element could be discerned, what then could we say of it in terms of statements understandable to the situation and yet in relation to the generic? Forcing is a 'what if' construction asking what if there existed such and such a multiple, with such and such a name that is retroactively justified through subjective fidelity to the event? Can we not be satisfied with that? The technical answer, which may not necessarily interest the general reader, is no, because as regards ontology, remember the generic is an element of ontology but not of logic as such, there is no temporality. Thus, the retroactive or retrospective element of forcing, that there will have been a reason to assume that a generic multiple exists, is meaningless and unacceptable to the very forces which are our basis for saying that forcing should be included in the line-up of new philosophical procedures. Then again it would perhaps suffice to say that, faced with this issue, Cohen treats the indiscernible in terms of the undecidable so as to bypass the problem of the future anterior typical of forcing within the situation. So we could say forcing can be justified in the future anterior in the situation, but in terms of undecidability as regards mathematical ontology.

There are three reasons, I feel, why this compels us to spend a little more time on forcing before we depart from *Being and Event*. The first is that undecidability heavily implicates indifference. The second is that it is the subject who 'forces' in Badiou's philosophy not, we should say, in mathematics. To fully understand the subject we must know what they are doing when they force. And third, in that forcing concerns the possibility of ontology establishing a relation that is formally traceable within the world such as it is, obviously forcing exists at the sticking point of all Western ideas on ontology since the dawn of philosophy: How does Being as such relate to multiple beings in the world? The holy grail here is a means of thinking both about a generic multiple which can be controlled from the situation by being a pair of condition and understandable statement in the language of the situation, and yet which can have its veracity confirmed in relation to the generic extension. So, forcing is finding a statement that is constructible and yet true generically in terms of ontology. Or, to put it the other way around, there is a two-way transitive equivalence in forcing between situation and generic extension: 'any veracity in the extension will allow itself to be *conditioned* in the situation' (BE 411). This double economy results in a particular inhabitant of a situation, this will come to be the subject, being such that 'although an inhabitant of the situation does not know anything of the indiscernible, and so of the extension, she is capable of thinking that the belonging of such a condition to a generic description is equivalent to the veracity of such a statement within that extension … she forces veracity at the point of the indiscernible' (BE 411). And she can do all this from within the quasi-complete situation primarily because said situation reflects the axioms of ontology itself. Simply put, a subject combines faith in the hypothetical existence of an indiscernible within the situation, with a double proof. The first is the confidence that if such an element exists it would also be consistent as regards all other multiples. And second, if it exists, through the process of enquiry and decision, its existence can be shown retroactively to be the case through the process of naming, connection, testing and so on.

Putting aside the retroactive element we are left with the question: Can we systematically, from within the operations available to set theory, demonstrate the equivalence between a generic statement and a generic situation so that any statement pertaining to ♀ is meaningful to the situation, and any generic situation ♀ is expressible within the language of the situation? I will not detain the reader here except to say that, to Badiou's satisfaction, it is the case that this equivalence can be proven entirely from within the atemporality of set theory alone. All of this occurs, yet again, through the foundation of ordinal rank in relation to the void. This leads to one of three outcomes. That the formula we are considering, Badiou calls it λ, is veridical in *any* extension *S*(♀) because it is forced by the void alone which is always included. It is not veridical. Or it is veridical in some extensions of *S*(♀) but not others. This last point is the undecidability of the generic. Thus, the third situation is undecidable in the negative because although we know it belongs to the situation it is generic meaning we can't stipulate where or how it belongs. Yet it is also undecidable in the positive in that it is the only situation where a generic situation can generate novelty, and because its undecidability is actually discernible within set theory even if it is not decidable. In saying the element is undecidable one asserts that it can be veridical, it is just that from within the situation we cannot discern how it will be.

One important implication of this pertains to the nature of novelty as we have been debating it in terms of indifference. Does indifference equip the event to change elements within a situation only, or can it be the basis for changing the situation as such? Badiou definitively comes down in favour of the former when he shows that if the generic extension of a quasi-complete situation can be forced due to the axioms of ontology, this means that it is also quasi-complete in itself. For example, 'insofar as the generic extension is obtained through the addition of an indiscernible, generic, anonymous part, it is not such that we can, on its basis, discern invisible characteristics of the fundamental situation' (BE 417). There is, it transpires, a clear limit to truths in terms of their reliance on indifferent qualities such as he delineates here. If they are indifferent they cannot be discerned in the situation. Instead, a forced truth can support supplementary veridical statements, but because the fidelity to these statements is within the situation itself, one can never negate the principles of consistency that pertain to all discernible statements. 'This is, moreover, why it is the truth *of* the situation, and not the absolute commencement of another. The subject, which is the forcing production of an indiscernible in the situation, cannot ruin the situation. What it can do is generate veridical statements that were previously undecidable' (BE 417). There is, it would appear, a price to be paid for finding a means of extending pure being as such into the realm of language so as to articulate a theory of change that is both evental and legible. In terms of the old adage that there is nothing new under the sun, forcing states that there are new statements possible under the sun, but only *if* they are under the sun. While the situation can be disrupted, as Badiou rather ominously states, whatever you do within the situation, one can never ruin the situation as such. This is, for all Badiou's statements to the contrary, a fundamentally Foucauldian and Deleuzian position. The long and the short of it is 'Genericity conserves the laws of consistency' (BE 417) or, for Badiou at least, indifference is indifference within situation, but never indifference of the situation as a whole.

In the following section Badiou extensively uses the law of non-contradiction to demonstrate his point that 'it is thus impossible for a veridical statement in *S*(♀) to ruin the supposed consistency of S, and finally *ST*' (BE 418). *ST* stands for set theory here. From this neat set of proofs, entirely dependent on the law of excluded middle, he is able to ascertain that either λ is a statement in ontology, or it is an undecidable statement in ontology 'that is, we can supplement the latter just as easily with λ as ~λ' (BE 419). The only moment, for a mathematician, that this issue matters, and for that matter for Badiou, is when it comes to the point of excess. Excess is the only point in the entirety of this debate between ontology and language that the subject can make any difference by considering the indiscernible as undecidable within a situation. In particular Badiou is concerned with his theorem that 'statist excess is without measure'.

If you recall, for the state to be able to claim to recount and thus structurally stabilize what has been presented in a situation it needs to open itself up to the void. One of the roles of constructivist thought, Badiou has explained, is to make sure that this excess is measured and minimized. To attain this result, the threat of the void to the stability of the excess, whose indeterminacy plants a virus in the absolute determinacy of the state, is negated powerfully by accepting the tenets of constructability. There is a displeasing symmetry here in that the state can only be consistent if it allows for the inconsistency of the void, just as the event can only create instability if it accepts the stability of the state. Badiou now shows, once more, that the limitations on the excess of the state by constructivist thought cannot contend with cardinal numbers as regards the infinite. In constructivist thought the threat of infinite cardinality is held in check by a simple formulation wherein the precise measure of Wα can be expressed. Badiou, however, suggests that 'if we find generic extensions *S*(♀) where, on the contrary, it is veridical that $p(\omega_\alpha)$ has other values as its cardinality, even values that are more or less indeterminate, then we will know that the problem of statist excess is undecidable within ontology' (BE 419). If we can show this then the whole purpose of constructivist thinking, which is to control the differentiation between belonging and inclusion in *S*, in fact to make it almost appear as if *S* is a mirror of the initial situation, would be under question. 'In this matter of the measure of excess, forcing via the indiscernible will establish the undecidability of what that measure is worth. There is errancy in quantity, and the Subject, who forces the undecidable in the place of the indiscernible, is the faithful process of that errancy' (BE 419–20).

In a nutshell, we encounter Badiou's politics of the subject. No change can occur in ontology. It would appear that nothing new can be included in the situation. Yet in ontology there is a great deal of generic indifference. Generic indifference as such is untenable within the situation. Yet because of ontology through the axiom of choice it is absolutely the case that singularities could exist, and because due to generic extension it is totally understandable to an inhabitant of a situation that if they did exist they would be consistent with the axioms of *S*, all it takes is for someone to believe in the event, which is impossible in ontology, and then speak of the event generically, which is impossible in the situation. To do this all the subject actually has to do is speak of the indiscernible, which has no existence in a situation, as an undecidable instead. This element, they say, which is currently indiscernible to the language of the situation, will come to have been discerned and so it is not indiscernible it is simply an element in *S* that has yet to be decided by set theory axioms. This is why the event's naming has to be compatible with *S*. It needs to exist in *S* due to its non-being in ontology otherwise no singularity would ever have an effect. More than this the effect of the event can only occur in a moment where *S* is weak, where it can admit of excess. There is no errancy of quality in *S* as all elements of *S* are differentiated due to quality (classification). The only gap in the entire system is to be found in errancy in quantity at the moment when the state tries to control the size of an unlimited situation (infinite cardinals) through aggressively tying inclusion to belonging. If the state admits that there is an infinite amount that can be included, then they need to be able to measure that infinity, which they can. But as soon as they admit to the axiom of choice, which, remember, in constructible sets they do not, then there is an indeterminate number of infinitely included elements.

It is not vast numbers that the state fears, it is unaccounted for numbers, non-denumerable numbers. Thus the whole point of the subject is to force the state, and the entirety of being is primarily to support the possibility of an event so that a subject can force a truth that will disrupt the state. If, that is, you want your event to have political effects.

Badiou now meticulously proves that there is an indeterminate cardinal superior to ω_0 such that 'we will have thereby demonstrated the errancy of statist excess, it being quantitatively as large as one wishes' (BE 420). Yet, having demonstrated errancy, there still remains the question is this errancy a cardinal in the situation *S*? This is yet another reminder that what is possible in ontology is not necessarily attainable in the situation even though both are determined, in volume one of *Being and Event* at least, by the consistent laws of set theory. This requires further demonstrations which I will spare the reader. Suffice to say that not only is there a demonstration of indeterminate or indifferent excess, but such an excess can exist in a situation. The second point is particularly relevant when it comes to the question of the rarity of events. There is, after all, an infinite amount of indeterminacy in ontology. Then again, as regards every situation, there is the potential for an evental site, yet in the end there are very few events. I must concede that I am not sure of Badiou's reasoning for the intermittent scarcity of events. It appears primarily empirical, we can observe only a few events, but in these difficult final pages there also seems to be a mathematical reason as well, it is difficult to show that events can exist in a situation. In that events introduce measureless excess into a state then there are state-political reasons for limiting events. In that the job of the subject is technically difficult and requires blind faith, are we to assume a kind of 'psychological' reason for evental rarity? Not really, as subjects are not people.

There is, I would contend, an appeal to intuition here or better a double appeal, to justify the intermittency of events. The first request is that readers such as we are believe that there are events. Maths on its own cannot help the event in this regard. Someone must make the effort to assume the event because otherwise singularity in ontology and logic is 'impossible' and so, because neither discipline has a theory of the absolutely ostensive, said singularity will be given no space in the consistent axioms of either system. The second intuition is tied to the first. Given that events exist to people such as we are, and we are conscious of the interdictions of the state against their existence, then naturally events are few and far between. We can look to the avant-garde heroes of history and see that this is the case.

Yet without this communicable, empirical and intuitive relation to events, they exist but mostly they do not, one could approach the situation differently.

One could first say that events don't exist and that those historical traces are not the result of events at all. All truth is generated from knowledge, full stop. I think Badiou has a strong proof that this is not the case due to the axiom of choice. Or if they exist then we could say there are an infinite number, an actual infinite number. This is basically Deleuze's position. I cannot see a strong reason for the intermittency of events in this regard. The only thing that determines, actually, an event, are the rules of the communicability of the situation. This then appears to be Badiou's pact with the statist devil which, I think, also suits his own historical acsesis, Laruelle's main problem with the entire Maoist nature of Badiou's non-democratic philosophy.[10] Naturally, the state will not admit to many, if any, events. And by definition the subject needs to speak of the event in the language of the situation as we have seen. Thus, if the subject wants to believe in events, then generic extensions from ontology into the situation are the only means of their realizing their dreams. The state will try not to allow any of these, but just as the subject needs the state to speak of events, so the state needs the void to stabilize the actual innumerable nature of its being an infinite multiplicity of multiples. This, then, is as close as we can come to justify that events are rare and it is a constructivist argument in the end. Because events can only be spoken of as events due to the situation, then the situation can have some control over events and limit their occurrence to none, one, two or a few. The only way out of this would be to reject the situation as such. Yet to reject the situation means to give up on the proof of Being as stable due to the convocation of the void and the actual infinite, and Badiou's immortal observation that, given an absolutely stable Being, and a stable being of all beings, the void is necessary and this means that there is nothing to stop indeterminate choice. So better a rarity of events than no events at all.

Badiou concludes as follows: 'Statist excess is effectively revealed to be without any fixed measure; the cardinality of the set of parts ω_0 can surpass that of ω_0 in an arbitrary fashion. There is an essential undecidability, within the framework of the ideas of the multiple, of the quantity of multiples whose count-as-one is guaranteed by the state (the metastructure)' (BE 426). It has taken hundreds of pages to come to this conclusion but it is a highly significant one. Although perhaps Badiou is too rapid in transposing this mathematical/logical discovery to the actual world of real political occurrences, he does not, for example, do much work to show in detail how a state in ontology is the same as an actual state just because they have been given the same name by Badiou, still the point remains. If we take it to be the case that situations are constructible, and we might do so as an ordinary language theorist inspired by Wittgenstein, as a follower of constructivist logic represented by Dummett, as an advocate of Bodies Without Organs in Rhizomatic lines of flight after Deleuze, or as the totality of discourse and power after Foucault, then we must abide by the laws of constructability.[11] These are the axioms of constructible set theory or set theory without the axioms of foundation and choice. If we accept these axioms, then we must accept the wider set of axioms pertinent to Being as such. We must admit of being as such as it is only through being able to prove a stability within the infinite multiplicity of multiples that make up the constructible universe. If we accept the axioms of ontology as the basis for those of construction, then we cannot rule out the axioms of foundation and choice. This means three things. First, the indifferent nature of ontology will extend into the totally differentiated realm of constructed situations. Second, there is nothing that constructible set theory axioms can do to deny totally that if indifferent elements existed, they could be constructible. Third, that when it comes to the cardinality of the infinite sets of the constructible universe, there is no way to ban the existence of excess due to indeterminacy except the desire of the state to limit this excess. If limit on excess is nothing other than a matter of will, then a subject can exist who has the faith to will the possible existence of as yet indiscernible elements because within the situation they find themselves in they can be defined as being as undecidable as the excess of the infinite, indeterminate cardinals. The final proof, therefore, is a basic syllogism. Given constructability and ontological set theory stability, there is as little to stop a subject to truth speaking of an event as there is to justify the state refusing to do so.

From the indiscernible to the undecidable

This is the compact version. Now we should go over Badiou's thirteen-part unpacked version, for the record so to speak, and because in the end everything hangs on this issue of forcing the indeterminacy of state excess. So, we start with the quasi-complete denumerable situation. Everything comes from and operates within this situation. This leads to three conditions: partial order (some conditions are more precise than others), coherency (or compatibility) and what he calls liberty (incompatible dominants). These three conditions combined determine the two laws of a correct part of a situation, RD1 and RD2, to an inhabitant of that situation. Some of these correct parts will be generic because 'they avoid any coincidence with parts which are definable or constructible or discernible within the situation' (BE 427). In general, a generic part does not exist in a situation. So an inhabitant of a situation can have a concept of generic parts, but they can never possess them. She can, in other words, only believe in them. Yet names do exist and names allow the combination of conditions and other existent names so that one can work out the referential values of names via hypotheses concerning the generic part which itself cannot be known. This is actually composed constructively to name that which cannot be constructed. The law of constructability can easily be turned against it.

If you can fix a referential value to all the names which belong to the generic situation then this is called the generic extension; the only way that indiscernible elements, events, can be felt in a situation. This is definitely an extension. An important maxim, because the problem of communication between Being as such and beings in the world has remained unsolvable up to this point. The extensive nature of the generic can be proven by showing that each element of the situation has its own name, so called canonical names that 'are independent of the particularity of the supposed generic part' (BE 427). Independent of the generic, only available to ontology, means existent in the situation. The generic part, however, which can't be known by a situation, is an element of the generic extension all the same. 'Inexistent and indiscernible in the situation, it thus exists in the generic extension. However, it remains indiscernible therein' (BE 427). There is, in other words, a degree of asymmetry here which is fundamental to the division between belonging and inclusion that the entirety of all versions of set theory are based upon.

Through forcing you can define relations between conditions and formulas applied to names. So you can, if you will, force it to be the case that such a situation conforms to the quasi-complete situation's rules and thus it is constructible albeit generically and indifferently so. If you can force or show a statement to be veridical in a generic extension, then it can conform to the laws of the situation whilst originating from the generic part found only in ontology. So, forcing is the economy of the communication between ontology as such and the situation. This is an essential statement. By forcing you discover a profound affinity between the generic extension and the situation in general so that you realize that if the situation is quasi-complete, so too is the generic extension. They speak the same language, only one is rather vague in terms of what it is precisely they are speaking about. This is all very pally and amicable until the conversation turns to thornier matters: 'But certain statements which cannot be demonstrated in ontology, and whose veracity in the situation cannot be established, are veridical in the generic extension' (BE 428). Badiou specifically names the issue around the indeterminate, infinite cardinal. This is another way of saying in finding that ontology and the situation can speak the same language as regards hypotheticals, they discover that they admit to that which they cannot admit to: singular events which are indifferent and thus indiscernible. Badiou's sleight of hand here is rather pleasing. The two things ontology and the situation cannot speak of, are the two things they find they are speaking of, when they discover a common language around the generic small-talk of the conditions they both hold in common. This leads to Badiou's final maxim, maxim *l.*, 'One can thus force an *indiscernible* to the point that the extension in which it appears is such that an *undecidable* statement of ontology is veridical therein, thus decided' (BE 428). And if you *can* force, according to Badiou, 'someone' will force.

It is possible to sum up the totality of the study with the following sentence: 'This ultimate connection between the indiscernible and the undecidable is literally the trace of the being of the Subject in ontology' (BE 428). Not least because central to the formulation of the being of the subject are two aspects of indifference. It is to Badiou's immense credit that he is able to differentiate that which is indiscernible from that which is undecidable. The indiscernible is truly indifferent and can never exist in the world. This is our own maxim that we have been true to all the way through our study. Indifference 'exists' in ontology but it inexists in the world which is, by definition, differential. Yet the undecidable, although generic and thus at this moment indifferent, will be differentiated. The undecidable is the last chapter in the history of indifference and the only way that indifference can be extended into a world of difference. To be allowed to speak as it were, indifference must accept the hemlock of a future decision.

Clearly indifference is not Badiou's major concern in the closing comments, rather he wants to leave us with a finally completed portrait of the being of the subject and we should respect that. The being of a subject, then, is the trace in ontology of a connection between what is indiscernible (ontology) and what is undecidable (language). This connection occurs at the point of statist excess or proof that indeterminate values for infinite cardinals exist. This shows, Badiou says, the failure of ontology to '*close* the measureless chasm between belonging and inclusion' (BE 428). The connection itself exists because there is a 'textual interference between what is sayable of being-qua-being and the non-being in which the Subject originates' (BE 429). Badiou is careful here. The chasm between the sayable and the inexistent spanning ontology and language is, as we have repeatedly said, one of the two areas that philosophy has always foundered. In that Badiou has solved the problem of stability by negating infinite regress with the void and proving actual infinity with indifferent multiplicity, he must then solve the other great pressing problem of metaphysics, some mode of articulation between the non-relational worlds of pure abstract being and the actual world.

His solution is odd. Set theory basically solves this problem through the state, but Badiou shows that the state's attempt to control excess through ontological structure is weak. Yet the very thing which makes the economy of ontology and logic weak, the existent indeterminate, is ultimately what allows them to interfere with each other through the subject's faith. Badiou, however, does not then say that said interference proves communication across structure and language, but rather that the subject 'must *be capable* of being'. The event, banned from ontology, 'returns in the mode according to which the undecidable can only be decided therein by forcing veracity from the standpoint of the indiscernible' (BE 429), or the only being which a truth is capable of is due to its 'indiscernible inclusions'. And this simply means that it allows the effects of said inclusions, which were up to this point suspended (not decided upon), to be 'retroactively pronounced such that a discourse gathers them together' without said gathering being annexed to the encyclopaedia. This has the further implication that the totality of that which the Subject is in its being 'can be identified in the trace at the jointure of the indiscernible and the undecidable', a jointure mathematicians have named, with just the right degree of blind perspicacity for our purposes, forcing.

All of which combined shows that the impasse of being, 'which causes the quantitative excess of the state to err without measure, is in truth the pass of the Subject' (BE 429). This is because 'A subject alone possesses the capacity of indiscernment ... It is thus assured that the impasse of being is the point at which a Subject convokes itself to a decision, because at least one multiple, subtracted from the language, proposes to fidelity and to the names induced by a supernumerary nomination the possibility of a decision without concept' (BE 429). For us, this is the last breath of indifference in Badiou's work but what a perfectly poisonous exhalation it is. He says: 'it is not impossible to decide ... everything that a journey of enquiry and thought circumscribes of the undecidable' (BE 429). In that the undecidable is indifferent, albeit indifference under a suspended death sentence, it is clear that right up to the last Badiou's ontology is a philosophy of indifference. The very phrase 'decision without concept' sums up the entire, complex and counter-intuitive logic of indifferential reasoning. It states that one can decide, one can deduce, without a specific concept in play ... yet. Thus, in set theory, we saw that one could gather and separate elements that were quality neutral because they were relationally quantifiable. Then we observed that in fact the relationality itself was reducible to pure quantity. Now we realize that in the world of situations one can make decisions as regards indifferent elements, and decisions mean differentiation. One can make these decisions not due to quantity, but actually due to the impossibility of determining quantity when it comes to certain infinite groupings of 'large cardinals' as they are called.

The problem here is almost the opposite. In ontology, there is no need for quality because there is relation which itself is reducible to quantity. In language, there is nothing but quality and relationality so instead the indifference of quantity means one can decide without a concept, or one can speak of an indifferent quality. There is, in a sense, nothing in the history of Western thought more paradoxical, more despised than indifferent quality and that is why decision is promised here. Yet whilst one might say, from the perspective of difference, indifference in the language of the world will always be decided, and so indifference negated, one can just as easily say even in the world of relational quality, due to undecidability, there is always indifference.

Conclusion (Meditations Thirty-six and Thirty-seven)

The final meditation is important in terms of time and place. Badiou, writing in the late-1980s, in France, works hard to vouchsafe his concept of the subject against opposing and dominant French voices: Descartes and Lacan. We don't need to spend any effort on this attempt. To a large degree, the difference between Badiou's subject and that of the history of Western thought is, like the fact that we can now talk freely of the subject, proven by the success of Badiou's work and the brilliant commentary it has attracted. As regards Badiou's debts to Lacan, these are widely commented on so again there is freedom for us to not dwell on those. Certainly, Badiou speaks again here of indifference of truth as regards language as a whole, 'since its procedure is generic inasmuch as it *avoids* the entire encyclopaedic grasp of judgements' (BE 433), but again it is now well-known that Badiou poses perhaps the only credible alternative to the double linguistic turn of analytical and continental philosophy, and it is also a point he returns to with fervour in the opening pages of *Logics of Worlds*. Even the three-part conclusion of the book reads a little like old news now as regards the impact Badiou's work has already had. I am certain there are many readers who still may not accept these propositions, but the fact the Badiou can credibly make them seems common currency. So we will not end on his tri-partite conclusion:

that the ontological basis of philosophy since the Greeks can, contra-Heidegger, be investigated through mathematical ontology; that logico-mathematical procedures since Frege and Cantor with be the new language of such a revolution in thought; and that finally we can keep the subject and speak of it through considering the generic procedures of aesthetics, science, politics and love.

Instead, we will close with the final words of the penultimate meditation, a meditation kept from being the capstone one suspects for strategic reasons, it is simply too technical and challenging, too bitter in a sense, to end the banquet on. Yet it is here that we find the thesis of the book, rather than speculations over its possible effects. Badiou is speaking of truth presented through torsion. In one sense, in speaking of the subject to truth Badiou is really speaking of how to undermine the state. In another sense, in speaking of the excess of the state he is showing that there is such a thing as truth by virtue of the subject's faith. These two are not mutually exclusive conceptually, in fact they present Badiou's specific sense of truth as being something political or revolutionary, and every revolution resulting in some conceptual truths. For Badiou, truth or veracity has two sources. The first is being 'which multiplies the infinite knowledge of the pure multiple'. The second is the event 'in which a truth originates, itself multiplying incalculable veracities' (BE 430). Here is why the book is called *Being and Event*, and does not mention the subject. In that Badiou's quarry is a legitimate and rational theory of truth against the depredations of the philosophies of difference in one tradition, and constructivism in the other, the subject only serves that. As he says of it 'Situated in being, subjective emergence forces the event to decide the true of the situation' (BE 430). This, and only this, is the role of the subject.

The wider implications of this assertion of the event gift to us perhaps the strap-line for the whole book: 'There are not only significations, or interpretations. There are truths, also' (BE 430), a position restated significantly at the beginning of *Logics of Worlds* with the important modification that there are only languages and *bodies*, except there are truths.[12] The theory of bodies is an essential innovation added to the two-volume project not covered in volume one. These truths however are 'practical', by which read legible and provable. They are in part subtracted from language through indiscernibility, and in part subtracted from Ideas through undecidability. So, if truth is subtractive it is doubly so due to two differing aspects of indifference. Such a truth needs the 'presentative support of the multiple' to be in some ways material and also legible. Yet this alone is not enough for truth. Rather such a support is the downgrading of a truth into knowledge without the ultra-one of the event. Combine Being and event and: 'The result is that it *forces decision*' (BE 430), or negates, at last, indifference. When it does so the subject is wedded to force, which perhaps is why so many take it to be by definition political. As Badiou says: 'Every subject passes in force, at a point where language fails, and where the idea is interrupted'. But this violence named force is not quite all it appears to be as we saw. Force is not power. It is rather an insistence, a promise, an as yet undifferentiated speech act or oath. Anyway, whatever the nature of this 'force', what it allows the subject to 'open upon', he is purposefully using Heideggerian terms here against Heidegger, 'is an un-measure in which to measure itself' (BE 430), or the discipline of the conditions of measurement, set theory, without committing to any particular rank (ordinal) or size (cardinal). All of this because 'the void, originally, was summoned' (BE 430).

Although it is some time since we considered the void in relation to being and the state, it must never be forgotten that the entre edifice of ontological set theory and constructible set theory is the subtraction of the void from presentation and its inexistence in representation. So that while it feels Lacanian to say, in the last words of the meditation, 'The being of the Subject is to be symptom-(*of*-) being' (BE 430), what Badiou is actually saying is that you will know of the being of the subject because it is symptomatic of being as such in the manner we commenced with analysing. If being as such is that which is subtracted so that the pure multiplicity of multiples can be founded up to an infinite point that needs no totality, so the subject which exists entirely in this constructible universe of multiples is the concomitant result of that subtraction. You will know being by its not being in the world, and you will know the subject by its not being in ontology. The two elements co-found each other in a profoundly subtractive fashion that is impossible unless you concede that being is, by definition, indifferent.

Bridge: from *Being and Event* to *Logics of Worlds*

Badiou's approach, a procedural consistency applied to forms of true inconsistency, Being as the not-one or event as truly singular, solves a profound problem for philosophy: the impossibility of articulating the One and the many such that they remain intrinsically what they ought to be. Historically, exposure of the One to the many had ruined its stability, while the corralling of the many under the One negated the quality of change and variety that typifies the many. Now, after Badiou, the stability of the One is due to the many or multiple while the problem of change has escaped from multiplicity as such and been subcontracted to the agencies of the event. Badiou's solution certainly sounds ingenious but does it not come too late? After all, we are still left with the problem of proving the One or Being to be 'consistent' against a dual tradition that says Being is either in-withdrawal or was simply a semantic confusion. And it leaves this other difference, the event, a difference outside the intelligible terms of philosophy, in that it is absolutely different without any reference to Being whatsoever (meaning it cannot be a being in the world either), hanging.

In truth, Badiou could have left the problem of difference to one side. Proving that ontology is rationally, formally and operationally consistent without question was ever the quest of our community. Turning to difference after Hegel was a gesture of desperation when the parts of Being couldn't be made to fit. Who needs radical difference if we have an operative theory of Being and beings?

The term 'need' is relevant here because real difference, the event, only comes to those who desire it and decide on its existence; Badiou is resolutely one of those as are his followers. Badiou wishes to prove the perfectly consistent, absolutely stable reciprocity of Being and beings so that he can argue the existence of at least one real, inconsistent difference. He realizes that he cannot define difference from within identity, yet all those who have tried to do otherwise have concluded that even if difference precedes identity, without identity-formations of some stability, differences are imperceptible as truly singular. Yet the whole point of Badiou's ontology is to define a difference that is truly, radically, singular and different. Whereas philosophers of difference proposed this pathway by debilitating the strangle-hold of stable identity structures on how we conceive of beings, what Badiou realizes is that the opposite path is requisite. Difference did not fail to negate identity because identity was too strong, but because it was too weak. In that difference is a co-component of identity, not its secret or despised other as our 'Continental' community have tended to propose, to strengthen difference you need to make identity unassailable. This is the simple thesis of the Being and Event project, stretching over twenty years and two daunting tomes. What the two books in tandem assert is that if there can be proven to be such a thing as a formally consistent theory of ontology that does not succumb to the traditional aporias of foundation, upper limit and internal coherence, then it can be shown that there can be such a thing as real change if you choose to decide on it first, then consistently pursue its consequences ever after. Accept that there is a stable theory of Being determined entirely by what it is composed of, beings in worlds, and against that absolute stability ask once more the question, is there such a thing as real difference?

Prove being so that you can decide on events in such a way as they are formalizable in terms of their effects on already existent states.[13]

In this way, going against our own contention somewhat, Badiou is a philosopher of difference, *the* philosopher of difference, the *last* philosopher of difference. Like his forebears he is not concerned with the stability of states per se, he thinks this self-evident and total. Instead, he is interested in Being only in as much as, in determining a true stability of identity, it allows that if you can prove that there is something, relative to that consistency, which presents itself to a world and yet which is inconsistent relative to it, then that is indubitably real change, an event. Again, in a way he takes this as self-evident. He sees events all around, not many but they are tangible to him and unquestionable, just as he sees stable states. That, however, is not enough for everyone. Followers of Badiou, as he said in conversation with Bosteels, are already sold on innumerable philosophies of various kinds of event. If the truth of events is to be accepted it must be proven beyond all reasonable doubt to all kinds of thinkers if philosophy is to remain a communicable pursuit, and to do this it must be asserted indubitably that something exists in a stable world which is totally inconsistent with that world. Define that world entirely due to relation, as Badiou eventually does in *Logics*, and what you are saying is there exist, on occasions, units presented in ontologically 'consistent' and logically stable situations which are, due to the fact that consistency is defined as relation, nonrelational. That the consistency of the relational world, which is all that exists, is entirely dependent on a consistent ontology, and the fact that the possibility of an event is broached first by events being generated by and ejected from pure ontology, and second that their reality in worlds, is based on certain foundational laws of ontology, because of being it can be said that there are events.

Badiou has two bites at this delectable, but oft-times poisoned, cherry. In *Being and Event* he establishes a consistent theory of ontology through the application of the axioms of mathematical set theory. From this unquestionable foundation he proves that truly singular, real and radical differences cannot be disproved. If being is consistent then events are at least possible. However, in that events contravene one of the basic laws of ontology, they are self-belonging or impredicative, it is impossible to accept events at an ontological level and maintain the rational consistency of that level due to the axioms of set theory. Badiou then has to participate in an emergence theory of the event. The non-presentable nature of events means that they emerge from the real of Being where they in-exist, into the realm of existence or states as Badiou calls them in the first volume, in an apparently non-causal fashion. It is now widely regarded that Badiou failed to achieve a credible theory of the event in this manner.

In the second volume of the project he comes at the problem in a different way. In *Being and Event,* the task was: Can you even speak of real change as events? A consistent ontology meant you could at least not dismiss them, but it also stated that you could not compose them within the ontological field. In *Logics of Worlds*, Badiou accepts that events are impossible ontologically and the role of ontology relative to events alters. Ontology still shows that there could be events, but now its main role is to provide a stable basis for situations, states or worlds as he calls them from 2006 onwards. In a way, the argument is the same, prove an absolute consistency such that a real inconsistency can occur. Only now it is worlds that are proven to be absolutely consistent, because it is only in terms of existence, not being as such, that events can in any way present themselves. Banned from ontology yet given a chance for existence thanks to the axioms of being, events must occur in worlds.

The consistency of worlds is hard to define. Badiou needs to combine set theory mathematics with category theory mathematics and use both classical and intuitionist logic to even be able to say that worlds exist which are consistent. Further, the ban on self-belonging in terms of being is also in place when it comes to beings in worlds. More than this a new limitation is imposed. Existence in a world is defined in terms of relationality, and events, to be truly different, have to be totally nonrelational. An event, on this reading, must be an element expelled from pure being, somehow present in a world, but in such a way that it contravenes the basic rules of what it means to be present in a world. Even worse, every world is made up of differences, and is composed of an infinite number of those. This means that worlds are defined by a stability of differential changes. If an event were to occur there, how would one even be able to be sure it was an event, and not just another predictably new thing?

Badiou's is, in effect, an absolutely compelling story of intellectual derring-do. In volume one, events are proven to be possible but they can find no place in the world. In volume two, they are presented to worlds, but in such a way as their existence seems to be impossible. Will our hero succeed in convincing us of his conviction that at least one event 'exists'? To do so he will have to persuade us of the following propositions, to make events transmissible, intelligible and communicable amongst us.

- Being inconsists in a formally consistent manner.
- Events are possible, in that if being is as set theory describes it, events cannot be disproved.
- Events are impossible to speak of ontologically, so if they exist they must present themselves in actual situations, states and worlds.

- For events to exist in worlds, again worlds have to be proven to be absolutely consistent even though they are definable primarily in terms of constant change and are radically, infinitely non-completable.
- For worlds to be consistent the beings of those worlds have to be defined as relationally consistent in terms of their difference.
- Worlds are both determined by constant change and relational difference, so any event must be a different kind of change due to a nonrelationality, a proposition which appears to contravene the very idea of events as 'different'.
- In that events are absolutely forbidden to the formal method of ontology, set theory, they can only exist in worlds. For events to exist in worlds, worlds have to be as stable as ontology. For them to be stabilized they must succumb to the laws of ontology, as well as the ban on self-belonging, and they must impose a new law, that of relationality. Yet events, we will come to see, will be defined as self-belonging multiples, which are fundamentally nonrelational. How can this circle be squared?

Basically, Badiou has to prove a stability to make events possible then prove another stability to make events actual. It is an immense task but, perhaps, not an insurmountable one. We will follow him step by step in the second volume of our study having secured a great deal of territory in this book. First, we proved ontology to be formally stable due to the axiom being is-not. Second, we showed that a stable being allows for a stable collection of beings; call these sets of multiples. Third, it was made clear that to attain stable states, a degree of excess along with at least one inconsistent element must be accepted. Fourth, we used the combination of excess and inconsistency as a platform for the construction of a theory of the event. Finally, fifth, we attempted to prove the existence of events in actual states represented by real 'subjects' intentionally behaving in a different way because of their faith in the fact that a new truth has been created.

At every stage it was apparent that these immense achievements made by Badiou were only possible due to indifference. So, we can say that first the axiom being is-not is made possible due to the fact that the void is in-different. Second, a stable collection of beings is due to the content neutral indifference of multiples. Third, the theory of excess was only possible due to the indifferential nature of sets as content-neutral collections. Fourth, no theory of the event is possible unless you accept it is nonrelational and content-neutral, hence doubly indifferent. Finally, fifth, no theory of the subject is possible without the process of forcing generic subsets and, as we saw, from top to bottom, generic subsets are indifferent.

When we began this study, we argued that it was because being was indifferent that there could be truly different events. What we discovered was that in fact indifference extends well beyond the confines of being, determining in essence what an event, a truth and a subject are. All the same, at the moment when a subject to truth starts to create a new world within a world by relationally building up from the event, or multiple which self-belongs, a new world composed of multiples that maximally belong to that first, impossible evental moment, it would seem inevitable that such worlds are differential, as all worlds are. Does this mean our philosophy of indifference, so powerful in fully explaining both being and event, will be useless as we turn to worlds and their relational differentiality? Or does communicability depend on indifference as I have argued elsewhere?[14] The reader will have to wait and see but I will furnish them with a clue. In that I sincerely believe that the epoch of the philosophy of difference came to a close between 1982 and 1988, with the publications of Agamben's *Language and Death,* Laruelle's *Philosophy of Difference* and, most significantly, Badiou's *Being and Event*, if *Logics of Worlds* is the masterpiece I have claimed it to be, then rest assured a large portion of its genius must be due to its being of its time, and its time, is unquestionably, our new age of indifference.

Notes

Introduction

1 Miguel de Beistegui, 'The Ontological Dispute: Badiou, Heidegger, and Deleuze', in Gabriel Riera, *Alain Badiou: Philosophy and its Conditions* (New York: SUNY Press, 2005), 48.
2 Jean-Francois Lyotard, *The Differend*, trans. Georges Van Den Abbeele (Minneapolis: University of Minnesota Press, 1988).
3 Alain Badiou, 'Can Change be Thought? A Dialogue with Alain Badiou', in Riera, *Alain Badiou*, 247.
4 Ibid.
5 See Jean-Toussaint Desanti, 'Some Remarks on the Intrinsic Ontology of Alain Badiou', in Peter Hallward (ed.), *Think Again: Alain Badiou and the Future of Philosophy* (London: Continuum, 2004), 59–66.
6 Several of the monographs published on Badiou concern themselves in detail with *Theory of the Subject*. In particular, see Bruno Bosteels, *Badiou and Politics* (Durham: Duke, 2011), 1–156; Ed Pluth, *Alain Badiou* (London: Polity, 2010), 104–27; Oliver Feltham, *Alain Badiou: Live Theory* (London: Continuum, 2008), 32–83; and Jason Barker, *Alain Badiou: A Critical Introduction* (London: Pluto Press, 2002), 13–38.
7 Alain Badiou, 'Can Change be Thought?' in Riera, *Alain Badiou*, 252.
8 Ibid., 252–3.
9 Ibid., 237.
10 For more on this see Brian Anthony Smith, 'The Limits of the Subject in Badiou's Being and Event', in Paul Ashton, A.J. Bartlett and Justin Clemens (eds), *The Praxis of Alain Badiou* (Melbourne: re:press, 2006), 75.
11 The best overall analysis of Badiou's dialectic of consistency and inconsistency is Alex Ling, 'Ontology' in A.J. Bartlett and Justin Clemens (eds), *Alain Badiou: Key Concepts* (Durham: Acumen Press, 2010), 49–51. Although we don't have time here to consider them, there are intriguing discussions of quasi-consistency in Riera, 'Alain Badiou: The Event of Thinking', in Riera, *Alain Badiou* 7, of the possibility that inconsistency precedes consistency in Ling, 'Ontology' in Bartlett and Clemens (eds), *Alain Badiou* 49; and finally of the lack of direct movement from inconsistency to consistency in Sam Gillespie, *The Mathematics of Novelty: Badiou's Minimalist Metaphysics* (Melbourne: re:press, 2008), 15.

12 Technically there is no proof of this. Being is-not is an existential axiom. That said, the amount of 'provable' operations it allows is so immense, so transmissible, that we will take this to be as close as a proof one can get in this world.

13 As we won't say much about poetry as we progress it is worth noting that while we choose to present this debate in terms of ontology and mathematics, the same discussion is underway, according to Badiou, between the poetry of Pessoa which presents nothing but 'multiple singularities', and Mallarmé who makes the mistake of pursuing a concept of the One (Alain Badiou, *Handbook of Inaesthetics,* trans. Alberto Toscano (Stanford CA: Stanford University Press, 2005)), 44. Hereafter referred to as HI.

There are now several excellent studies of Badiou's relation to the poetic 'condition' as he calls it. Premier of these are Jean-Jacques Lecercle, *Badiou and Deleuze Read Literature* (Edinburgh: Edinburgh University Press, 2010) and Andrew Gibson, *Beckett and Badiou: The Pathos of Intermittency* (Oxford: Oxford University Press, 2006), but mention should also be given to Elie During, 'Art' in Bartlett and Clemens (eds), *Alain Badiou* 82–93; Gabriel Riera, 'For an "Ethics of Mystery": Philosophy and the Poem' in Riera, *Alain Badiou* 61–86; Jean Michel Rabaté, 'Unbreakable B's: From Beckett and Badiou to the Bitter End of Affirmative Ethics' in Riera, *Alain Badiou* 87–108; Pierre Machery, 'The Mallarmé of Alain Badiou' in Riera, *Alain Badiou* 109–16; and Jacques Rancière, 'Aesthetics, Inaesthetics, Anti-Aesthetics', in Hallward, *Think* 218–31.

14 Badiou begins to speak of nonrelationality in Marxist terms in Alain Badiou, *Theory of the Subject*, trans. Bruno Bosteels (London: Continuum, 2009), 128. Hereafter referred to as TS. And although this falls away in the later work it reveals a fascinating insight into the gestation of this all-important concept. That said nonrelationality is not entirely accepted amongst the Badiou community. Significantly, Peter Hallward does not accept nonrelationality in its full sense in Badiou's work. See Peter Hallward, 'Introduction: Consequences of Abstraction' in Hallward, *Think* 13–15. While Bensaïd negatively terms nonrelationality a 'miracle', in Daniel Bensaïd, 'Alain Badiou and the Miracle of the Event', in Hallward, *Think* 98, perhaps influencing Hallward's own position to some degree.

15 See G.W.F. Hegel, *Phenomenology of Spirit*, trans A.V. Miller (Oxford: Oxford University Press, 1977), 58–103.

16 See Badiou, *Theory of the Subject* 6 and 8. For more on this see Barker, *Alain Badiou* 151 and Lorenzo Chiesa, 'Count-as-One, Forming-into-One, Unary Trait, S_1', in Ashton, Bartlett and Clemens (eds), *The Praxis of Alain Badiou* 158–62. See also G.W.F. Hegel, *Science of Logic,* trans. A.V. Miller (Amherst NY: Humanity Books, 1969), 825; Agamben's consideration of this crucial passage in Giorgio Agamben, *Potentialities,* trans. Daniel Heller-Roazen (Stanford CA: Stanford University Press, 1999), 116–18; and my interpretation of this exchange in William Watkin, *Agamben and Indifference* (London: Rowman & Littlefield International, 2015), 59–61.

17 It is astonishing how much has been written on Badiou's relation to Deleuze in such a short time. The readings of Badiou and Deleuze in this work are our own and are determined by a specific reading of Deleuze as rejecting indifference to establish his philosophy of difference influenced by Badiou's critical reading of Deleuze, especially in terms of the ubiquity and univocity of the event presented in Alain Badiou, *Deleuze: The Clamor of Being*, trans. Louise Burchill (Minneapolis: University of Minnesota Press, 2000), 19–30. Hereafter referred to as DCB. That said, I am naturally indebted to the various studies now available starting with Roffe's ground-breaking work, Jon Roffe, *Badiou's Deleuze* (Durham: Acumen Press, 2012), along with the innovations of A.J. Bartlett, Justin Clemens and Jon Roffe, *Lacan Deleuze Badiou* (Edinburgh: Edinburgh University Press, 2014); Jean-Jacques Lecercle's indispensable *Badiou*; John Mullarkey, 'Deleuze' in Bartlett and Clemens (eds), *Alain Badiou* 168–75; Barker, *Alain Badiou* 111–29; Todd May, 'Badiou and Deleuze on the One and the Many' in Hallward, *Think* 67–76; Daniel W. Smith, 'Badiou and Deleuze on the Ontology of Mathematics' in Hallward, *Think* 77–93; and Gillespie, *The Mathematics of Novelty* 1–24.

18 See Gilles Deleuze, *Difference and Repetition,* trans. Paul Patton (London: The Athlone Press, 1994), 28–69.

19 Badiou speaks negatively of this kind of indifference in relation to the anarchy of the Lacanian imaginary in TS 301. By the time of *Being and Event* his attitude towards quality-neutral indifference has altered dramatically and irrevocably.

20 It would be interesting to compare this history of indifferent difference with Badiou's own typology of pure difference using three different schema of representing strings of a's in *Theory of the Subject*: 'Each atom must be regarded as being . . . itself, that is, the indifferent "a" distinguished only by its place' (TS 70). This is the beginning of the more sustained presentation of indifferent multiples in terms of indifferent ranking that typifies *Being and Event*'s ontology.

21 See Badiou's summary of its importance in Alain Badiou, *Briefings on Existence: A Short Treatise on Transitory Ontology,* trans. Norman Madarasz (Albany, NY: SUNY Press, 2006), 41–2. Hereafter referred to as BOE.

22 In many ways, Zermelo-Fraenkel axiomatic set theory is a negation of central aspects of Cantor's work, not least his definition of the set as such and his pursuit of what came to be called the Continuum Hypothesis. For more on the complex history of set theory and Badiou's perhaps abbreviated presentation of it as the 'Cantor event' see Tzuchen Tho, 'What is Post-Cantorian Thought? Transfinitude and the Conditions if Philosophy' in Sean Bowden and Simon Duffy, *Badiou and Philosophy* (Edinburgh: Edinburgh University Press, 2012), 19–39; Desanti, 'Some Remarks', in Hallward, *Think* 60–1, Ling, 'Ontology' in Bartlett and Clemens (eds), *Alain Badiou* 51–4; Brian Anthony Smith, 'The Limits of the Subject', in Ashton, Bartlett and Clemens (eds), *The Praxis of Alain Badiou* 76–9; and Mary Tiles, *The Philosophy of Set Theory: A Historical*

Introduction to Cantor's Paradise (Mineola NY: Dover Publications, 1989), Chapters 5 and 6.

23 Badiou's choice of ZF+C as opposed to Gödel's seven-axiom constructible system is the main topic of discussion in *Being and Event* but it is interesting to note in the early work how he has to justify using set theory at all, compared with the relative merits of arithmetical combinatorics, see Alain Badiou *The Concept of the Model*, trans. Zachary Luke Fraser and Tzuchien Tho (Melbourne: re:press, 2007), 44–5. Hereafter referred to as CM.

24 Tho's point is well-made when he questions the overstating of dependency of Badiou's work on Cohen in Badiou's work itself (Tho, 'What is Post-Cantorian Thought?', in Bowden and Duffy, *Badiou and Philosophy* 26). When we speak of set theory then we are really considering a complex of names which Badiou comes to call the Cantor-Gödel-Cohen-Easton symptom in relation to choice for example (BE 280). Yet even this compound 'subject' omits the centrality of Zermelo and Fraenkel, as well as neglecting von Neumann and Dedekind, both of whom are essential components of Badiou's axiom 'ontology is mathematics'. Further, that Badiou rejects Gödel's constructible theory of sets does not mean he dispenses with Gödel whose incompleteness and completeness theories are a crucial component in the formalizing of the ZF+C model Badiou adheres to.

25 For a brilliant introduction to the philosophy of mathematics see Stewart Shapiro, *Thinking About Mathematics: The Philosophy of Mathematics* (Oxford: Oxford University Press, 2001).

26 Some have called this a two-multiple theory which is not entirely correct, rather, this is simply two different ways of looking at the same multiple. See Barker, *Alain Badiou* 45 for an alternate view and Gillespie, *The Mathematics of Novelty* 65.

27 I hope the reader will appreciate that God and the void are inexistent in profoundly different ways.

28 In the earlier work, Badiou sometimes speaks of retrospection in this regard, see for example TS 20–1 and CM 52–4. In *Theory of Subject* he also uses the term retroaction when he is differentiating his materialist dialectic from that of Hegel (TS 48) and when he is speaking of the effect of atom deviation as the result of the subtraction of the void (TS 62–3). However, perhaps the finest statement on retroaction can be found in the summary of knowledge in relation to novelty due to the subject that is, in effect, the thesis of the entirety of the book: 'I posit that there exists no intrinsic unknowable . . . what we did not know *before* was determined as a remainder of what has come to be known, at the crossover between the nameless movement through which the real appears as a problem and the retroaction, named knowledge, which provides the solution' (TS 201). For more on the relation of axioms to conditions see A.J Bartlett, 'Conditional Notes on a New Republic' in Ashton, Bartlett and Clemens (eds), *The Praxis of Alain Badiou* 212–19. For an

excellent summation of the centrality of the axiom for Badiou see Gibson, *Beckett* 49 and for a condensed yet revealing definition of the role of axiom to thinking see BOE 38.

29 It is important to note that if ontology is mathematics, Being as such is not mathematical. In addition, mathematics is just a discourse for Badiou who is not a mathematician nor an orthodox philosopher of mathematics either. See Ling, 'Ontology' in Bartlett and Clemens (eds), *Alain Badiou* 48 on this point.

30 On the relation between Plato and mathematics see Barker, *Alain Badiou* 46–9.

31 For more critical views on the axiom see Balibar's comparison of Badiou's axiomatic reasoning to that of Tarski in Etienne Balibar, 'The History of Truth: Alain Badiou in French Philosophy', in Hallward, *Think* 31; Ray Brassier, 'Nihil Unbound: Remarks on Subtractive Ontology and Thinking Capitalism', in Hallward, *Think* 55; and Daniel Smith, 'Badiou and Deleuze on the Ontology of Mathematics', in Hallward, *Think* 77–93.

32 The model is an important early vision of what came to be Badiou's use of set theory to solve the intractable problems of ontology. For more on this see Simon Duffy, 'Badiou's Platonism: The Mathematical Ideas of Post-Cantorian Set Theory', in Bowden and Duffy, *Badiou and Philosophy* 72; and Feltham, *Badiou* 22–6 who thinks of modelling in terms of retroaction and then interestingly contrasts the model to Badiou's later concept of the condition: Feltham, *Badiou* 132. See also Zachary Fraser, 'The Law of the Subject: Alain Badiou, Luitzen Brouwer and the Kripkean Analyses of Forcing and Heyting Calculus', in Ashton, Bartlett and Clemens (eds), *The Praxis of Alain Badiou* 43.

33 See also CM 42–3.

34 Badiou actually uses the term himself in Alain Badiou, *Metapolitics,* trans. Jason Barker (London: Verso, 2005), 18. Hereafter referred to as M. Indeed, the whole of the first chapter is effectively a discussion of communicability (M 10–25).

35 Michel Foucault, *The Archaeology of Knowledge*, trans. A.M. Sheridan Smith (London: Routledge, 1972), 86–7 and the discussion on 'The Second Moment of The Judgement of Taste' in Immanuel Kant, *The Critique of Judgement*, trans. James Creed Meredith (Oxford: Clarendon Press, 1952), 50–60.

36 For a consideration of the relation of communicability to the idea of community see Balibar in Hallward, *Think* 36 and Jean-Luc Nancy, 'Philosophy Without Conditions' in Hallward, *Think* 39.

37 See also Gillespie, *The Mathematics of Novelty* 87–90.

38 Alain Badiou, *Being and Event*, trans. Oliver Feltham (London: Continuum, 2005). Hereafter referred to as BE.

39 Communicability is also a central component of Badiou's rejection of what he calls anti-philosophy, in particular that of Wittgenstein which is so central to his rejection of constructivist, propositional, linguistic modes of thinking. See in particular Alain

Badiou, *Wittgenstein's Antiphilosophy*, trans. Bruno Bosteels (London: Verso, 2011), 107–21. Hereafter referred to as WA.

40 Although not completely. *Theory of the Subject* is a major philosophical work with implications for elements of *Being and Event* and, in particular, much to say about the articulation of the two volumes of the Being and Event project or the onto-logical as Badiou calls it. We will return to it repeatedly across both volumes of our study.

41 This is the subtitle of Hallward's foundational study of Badiou's work to which we are all deeply indebted.

1 Being: The One and the Multiple

1 For more on Badiou's interaction with Parmenides see Ling in Bartlett and Clemens (eds), *Alain Badou* 49; Hallward, *Badiou: A Subject to Truth* (Minneapolis: University of Minnesota Press, 2003), 310 and 322. Hereafter referred to as BST; and Christopher Norris, *Badiou's Being and Event* (London: Continuum, 2009), 37–41, 46–9 and 68–9. Hereafter referred to as BBE.

2 See TS 327 for a discussion of the ethics of the impasse: 'every rule is an im-passe of the real'.

3 It is also a major statement at the heart of *Theory of the Subject* which works 'To distinguish the One from the Whole' (TS 30).

4 François Laruelle, *The Anti-Badiou: On the Introduction of Maoism into Philosophy*, trans. Robin Mackay (London: Bloomsbury, 2013), xviii–xxxix.

5 Specifically, that they are not meaningful in any traditional version of sense and reference or use and mention. Rather, the meaning of one and many does not reside in what they refer to or to the cognitive benefits of their contextual expression, but solely in terms of what they do. In this fashion words like one and many are most similar to the idea of a mathematical function.

6 See Tiles, *The Philosophy of Set Theory*, Chapter 2.

7 For more on the use of retroaction in Badiou's work see Barker, *Alain Badiou* 46 and 59, Ling, 'Ontology' in Bartlett and Clemens (eds), *Alain Badiou* 49 and Gillespie, *The Mathematics of Novelty* 59.

8 Badiou sometimes uses the term retrospective instead of retroactive although in each case the meaning remains the same. We have decided to stick with retroactive because it highlights the operational nature of the process.

9 Considering the history of my work it may seem strange that I have studiously avoided interaction with Badiou's many, brilliant engagements with poetry, not least because he names poetry as one of only four conditions for philosophy. I have my reasons for this, one of which is the excellent commentary that already exists

covering Badiou and literature. However, I cannot resist here alerting the reader to Badiou's recuperation of the Heideggerian tradition of poiesis in his recasting of the poetry of the unpresented as the poetics of the void in the work of Rimbaud and Mallarmé (Alain Badiou, *Handbook of Inaesthetics,* trans. Alberto Toscano (Stanford CA: Stanford University Press, 2005)), 22.

10 For an alternative reading of the presentation of presentation in relation to Heideggerian presentative being and tautology see Barker, *Alain Badiou* 44.

2 Being: Separation, Void, Mark

1 See Norris, *Badiou's Being and Event* 37–52, also Ling, 'Ontology' in Bartlett and Clemens (eds), *Alain Badiou* 49.

2 Badiou is indebted to Lacan for this formulation of the 'pass' of the impasse. For more on this see Bosteels, *Badiou and Politics* 165–8 and Bartlett, Clemens and Roffe, *Lacan Deleuze Badiou* 127.

3 For Badiou's comparison of pure presentation and the 'there-is' of constructible ideas of substance, particularly those of Wittgenstein, see WA 97–9.

4 See Pluth, *Alain Badiou* 30–4, and A.J. Bartlett, 'Plato' in Bartlett and Clemens (eds), *Alain Badiou* 107–17.

5 For more on meta-ontology see Gabriel Riera 'Introduction' in Riera, *Alain Badiou* 8; Beistegui, 'The Ontological Dispute' in Riera, *Alain Badiou* 47; Barker, *Alain Badiou* 43; Ling, 'Ontology' in Bartlett and Clemens (eds), *Alain Badiou* 48, and Gillespie, *The Mathematics of Novelty* 49.

6 Michael Potter, *Set Theory and Its Philosophy* (Oxford: Oxford University Press, 2004), 3.

7 While it is true that no one, up to this point, has focused on the fundamental centrality of indifference to Badiou's ontology, as regards the indifference of set theory, gratifyingly, I am not alone in my contentions. See for example Barker, *Alain Badiou* 45–6, Pluth, *Alain Badiou* 40, Brian Anthony Smith, 'The Limits of the Subject', in Ashton, Bartlett and Clemens, (eds), *The Praxis of Alain Badiou* 76 and Ling, 'Ontology' in Bartlett and Clemens (eds), *Alain Badiou* 50. We have already seen that Norris concedes this point and Hallward also speaks regularly of indifference in Badiou even if he does not dwell on it, see for example Hallward, 'Introduction' in Hallward, *Think* on the indifference of mathematic abstraction (3), of pure multiplicity as such (3 and 5), quality neutrality (6), and indiscernibility (7). Perhaps the only critic of Badiou who does dwell on it is Etienne Balibar. Balibar's essay 'The History of Truth' considers the indifference of the generic (29), the impersonal indifference of the subject (32), of the truth procedure emerging out of indifference multiplicity (33), in relation to the indiscernible as generic (34), the neutrality of belonging (36), and finally the indifferent generic nature of fidelity (37).

8 Badiou also speaks of the 'separative efficacy of the axiom in general' marking set theory out as both axiomatically dependent on separation and also meta-mathematically a mode of separation due to indifferent modelling as well (CM 39). Many critics refer to this separative aspect of Badiou's work as a central component.
9 Indeed, as Ling points out ZF is not just an axiomatization of Cantor's work but also a negation of his basic concept as to what a set actually is (Ling, 'Ontology' in Bartlett and Clemens (eds), *Alain Badiou* 51–4).
10 Letter from Russell to Frege dated 16 June 1902 cited in Michael Beaney ed., *The Frege Reader* (Oxford: Blackwell, 1997), 253.
11 In Jacques Derrida, *Writing and Difference*, trans. Alan Bass (London: Routledge, 1978), 278–94.
12 See CM 16 and 19.
13 See Immanuel Kant, *Critique of Pure Reason,* trans. Marcus Weigelt (London: Penguin Books, 2007), ix. I am unaware of any other work that exists that points out that the incompleteness theory is behind Kant's critical philosophical project, but it clearly is.
14 See for example the essay 'Différance' in Jacques Derrida, *Margins of Philosophy*, trans. Alan Bass (London: Harvester Wheatsheaf, 1982), 1–28.
15 Martin Heidegger, *Identity and Difference*, trans. Joan Stambaugh (Chicago: University of Chicago Press, 1969), 70.
16 Badiou is very clear that he does not intend an ethics of alterity and indeed is a leading voice in its negation. See Alain Badiou, *Ethics: An Essay on the Understanding of Evil*, trans. Peter Hallward (London: Verso, 2001), 18–29. Hereafter referred to as E.
17 This is an important point as some critics, as we have noted, have tended to think of nonrelationality as partial rather than absolute. A related point is that if nonrelationality and events are nonrelational, which they are, then we can say that events can be prepared for. Certainly Badiou speaks of the quality of a subject of the wait in this regard, see for example Alain Badiou, *On Beckett,* eds. Nina Power and Alberto Toscano (Manchester: Clinamen Press, 2003), and Gibson, *Beckett.* But this is not the same as suggesting, as Hallward does, that events can be prepared for or in some way facilitated (Hallward, 'Introduction' in Hallward, *Think* 16–17). Worryingly, Hallward is not alone in this intuition. See also Ernesto Laclau, 'An Ethics of Militant Engagement' in Hallward, *Think* 124 and 135, and Balibar, The History of Truth' in Hallward, *Think* 32.
18 Another version can be found in the formula of appearance of appearance in Alain Badiou, *The Century*, trans. Alberto Toscano (Cambridge: Polity, 2007), 64. Hereafter referred to as TC.
19 The axiom of extension is already a central part of Badiou's materialism by the time of *Theory of the Subject*, see TS 192 and his comments on the isomorphism of set names.

20 Although we will not use this term again as Badiou prefers the more encompassing constructivism, it is perhaps the most well-known form of constructivism thanks to Wittgenstein. For a full understanding of Badiou's relation to Wittgenstein one needs to consult WA. For more on language games in general see Alain Badiou, *Infinite Thought*, trans. Oliver Feltham and Justin Clemens (London: Continuum, 2006), 35. Hereafter referred to as IT.
21 Meditation Six: Aristotle. As previously stated our intention is not to rehearse Badiou's readings in the history of philosophy. In the case of the sixth meditation on Aristotle for example, we can get away with saying, after Norris, that Aristotle famously ruled out the possibility of a vacuum in nature, that Badiou, redefining vacuum as void cannot agree with this position but, although physics has disproven Aristotle in any case, Badiou does not believe that Aristotle is talking about actual vacuums.
22 Alain Badiou, *Theoretical Writings*, ed. and trans. Ray Brassier and Alberto Toscano (London: Continuum, 2004). Hereafter referred to as TW.

3 Being and Excess

1 See an early use of formal excess in this manner in TS 166.
2 Badiou is already speaking in these terms before he has entirely adopted set theory. For example, in *Theory of the Subject* he defines atoms (later to be called multiples) as being 'identical with regard to the void' (TS 57), transposing this neutral atom into a discourse of strong versus weak difference. Weak difference becomes, in the later work, indifferent difference of quality neutral multiples. See also TS 61.
3 For more on this see Pluth, *Alain Badiou* 51–2 and Ling, 'Ontology' in Bartlett and Clemens (eds), *Alain Badiou* 52.
4 Critics often speak of Badiou's ontology as separative or partative and it is this section which confirms such a designation.
5 ZF was able to allow for one self-belonging multiple by off-setting the paradox that Russell found in Frege's original Axiom V, through the development of a larger and more sophisticated set of axioms. This is usually described as the transition from the naïve set theory of Cantor and Frege, to the axiomatic set theory we use today.
6 Or as he says in *Theory of the Subject*: 'The universe always contains more things than those it can name according to those things themselves' (TS 219).
7 Not everyone takes the empty set as read. For alternate views on this matter see Brian Anthony Smith, 'The Limits of the Subject', in Ashton, Bartlett and Clemens (eds), *The Praxis of Alain Badiou* 79 and Laclau, 'An Ethics of Militant Engagement', in Hallward, *Think* 126.
8 See Chiesa for a detailed consideration of these topics and their debt to Lacan in Ashton, 'Count-as-One', in Ashton, Bartlett and Clemens (eds), *The Praxis of Alain Badiou* 149–51.

9 Derrida, 'Signature Event Context' in *Margins* 309–30.

10 This procedure may be new to the reader but it is an absolutely mundane element of both mathematics and analytical philosophy of the last hundred years or more. The process of forming into one, although not called that usually, can be found in the work of Frege, von Neumann and others. What is unique here is Badiou's philosophical interpretation of the process.

11 The political implications of the term state have had extended ramifications which we will not explore here. For the curious, militant reader I suggest you commence with Bosteels, *Badiou and Politics*; Frank Ruda, *For Badiou: Idealism without Idealism* (Evanston IL: Northwestern University Press, 2015); Nina Power and Alberto Toscano, 'Politics' in Bartlett and Clemens (eds), *Alain Badiou* 94–106; Nina Power, 'Towards a New Political Subject? Badiou Between Marx and Althusser' in Bowden and Duffy, *Badiou and Philosophy* 157–76; and Nina Power, 'Towards an Anthropology of Infinitude: Badiou and the Political Subject' in Ashton, Bartlett and Clemens (eds), *The Praxis of Alain Badiou* 309–38. Finally, one should mull over the fifth chapter of *Metapolitics* 78–95.

12 Bosteels perhaps provides the best response to this question in Bosteels, *Badiou and Politics* 159.

13 For a clear differentiation between the two see Alain Badiou, *Mathematics of the Transcendental*, trans. A.J. Bartlett and Alex Ling (London: Bloomsbury, 2014), 13.

14 One of the attractions of set theory presented as a model is that it escapes the incompleteness theorem by being a strong example of what is called, in contrast, the completeness theorem: 'that of taking the inscriptions of a supposedly consistent theory for the model of the theory itself' (CM 64).

15 For Badiou's comments on the insistence on remainder in anti-philosophers see Badiou, WA 94.

16 See Sean Bowden, 'The Set-Theoretical Nature of Badiou's Ontology and Lautman's Dialectic of Problematic Ideas' in Bowden and Duffy, *Badiou and Philosophy* 46.

17 Pluth, *Alain Badiou* 54–5. See also Hallward, *Badiou* 99.

18 For a better understanding of Badiou's relation to Rancière see M 107–23 and Gibson, *Intermittency: The Concept of Historical Reason in Recent French Philosophy* (Edinburgh: Edinburgh University Press, 2012) 202–45.

4 Nature and Infinity

1 Badiou's relation to Spinoza is well-documented meaning we can pass over it with some haste. See, for example, Gillespie, *The Mathematics of Novelty* 25–42; Norris, *Badiou's Being and Event* 96–106; and Roffe, 'Spinoza' in Bartlett and Clemens (eds), *Alain Badiou* 118–27.

2 See Bowden, 'Set-Theoretical' in Bowden and Duffy, *Badiou and Philosophy* 48–9 on Badiou's decision to name the superstable metastructure.

3 Smith's distinction here between extensive presentation, or cardinal, and intensive presentation, or ordinal, is extremely useful as is his realization that this forms a kind of non-relation that is 'more of a non-determinate relation' (Brian Anthony Smith, 'The Limits of the Subject', in Ashton, Bartlett and Clemens (eds), *The Praxis of Alain Badiou* 75).

4 See CM 66.

5 Tho's comments in this regard are illuminating when he considers the means by which Cantor's thought allows for the 'unhinging of consistency and unity' due to the realm of the transfinite. It is, in other words, transfinite numbers which give the inspiration for the axiom being is-not defining being as a procedurally consistent inconsistency. See Tho, 'What is Post-Cantorian Thought?' in Bowden and Duffy, *Badiou and Philosophy* 28–9.

6 See Eric Steinhart, *More Precisely: The Math You Need To Do Philosophy* (Peterborough Ont.: Broadview Press, 2009), 151.

7 See CM 32.

8 Hallward, BST. (Minneapolis: University of Minnesota Press, 2003), 329.

9 See Alain Badiou, *Number and Numbers*, trans. Robin Mackay (Cambridge: Polity, 2008), 192. Hereafter referred to as NN. Also Tiles, *The Philosophy of Set Theory* 109; Steinhart, *More Precisely* 169; David Papineau, *Philosophical Devices: Proofs, Probabilities, and Sets* (Oxford: Oxford University Press, 2012), 25–6; and Gillespie, *The Mathematics of Novelty* 53–4 for different diagrammatic representations of the diagonal argument. What this argument shows is that real numbers cannot be put into a one-to-one correspondence with natural numbers meaning they are non-denumerable or inconsistent. What this proves is that infinite sets come in different sizes leading to Hilbert's famous remark on Cantor's trans-finite paradise. The trans-finite does not yet concern us, but will be relevant when we consider the relationship of non-denumerable or inconsistent infinities to the event. For our purposes here, however, what is significant is that once Cantor showed that infinities of different sizes, with different cardinals, existed, he was also able to show that they were at least partially ordered by using the concept of well-orderedness which we have been analysing, the remaining order being provided by the axiom of choice.

10 As does Badiou's clear assertion: 'When I say that all situations are infinite, it's an axiom. It is impossible to deduce this point. It is an axiomatic conviction, a modern conviction. I think it is better for thinking to say that all situations are infinite.' Badiou, *Infinite Thought,* 136.

11 This axiom is central to Badiou's thinking from an early stage, for example see TS 91.

12 See Stewart Shapiro 'The Philosophy of Mathematics and Its Logic: Introduction' in Stewart Shapiro (ed.), *The Oxford Handbook of Philosophy of Mathematics and Logic* (Oxford: Oxford University Press, 2005), 3–27 for a detailed presentation of these topics.
13 Sometimes the notation ℵ or aleph is used instead of ω.
14 Torsion is the interruption of a coherent, repetitious system by an element which can then become the component of another, alternatively coherent, repetitious system whose existence, however, makes the initial system incoherent. Torsion then is a mode in internal interruption best represented by the means by which, from the void set, through the simple operation of n + 1, facilitated by taking a set as a collection meaning an empty set can be counted as a one, a new serial set can be composed, as if out of nothing. That said, torsion is not the void set but rather the void set is what allows torsion as interruption to become torsion as coherent structure. Obviously, it is Badiou's first attempt at theorizing the formal possibilities of an event. Torsion is a dominant element of the system of *Theory of the Subject*, see TS 148–57. Perhaps the key quote is: 'repetition is thereby interrupted in favour of the advent of another coherence, from the standpoint of torsion, within the whole' (TS 153). For a detailed consideration of its significance see Bosteels, *Badiou and Politics* 107–9. Torsion of this order shares many similarities with the consideration of the event as self-belonging in *Logics of Worlds*, the event is a form of torsion in that it interrupts the successive coherence of a situation to develop another kind of or twist on (torsion means to twist) coherence. We will, then, return to it in the second volume of our study.
15 A lot has been written on Badiou's debt and critical distance from Lacan. See Bartlett, Clemens and Roffe, *Lacan, Deleuze, Badiou*; Feltham, *Alain Badiou* Chapter 2; Gillespie, *The Mathematics of Novelty* 104–21; Slavoj Zizek, 'From Purification to Subtraction: Badiou and the Real' in Hallward, *Think* 165–81; Joan Copjec, '*Gai Savoir Sera:* The Science of Love and the Insolence of Chance' in Riera, *Alain Badiou* 119–36; Juliet Flower MacCannell, 'Alain Badiou: Philosophical Outlaw' in Riera, *Alain Badiou* 137–84; A.J. Bartlett and Justin Clemens, 'Lacan' in Bartlett and Clemens (eds), *Alain Badiou* 155–67; Hallward, BST, 11–14 and last but certainly not least Chiesa's excellent essay 'Count-as-One, Forming-into-One, Unary Trait, S_1' in Ashton, Bartlett and Clemens (eds), *The Praxis of Alain Badiou.*
16 Ordinals are indifferent up to the point that they become retroactively differentiated by a re-count due to infinity.

5 The Event: History and Ultra-One

1 See Lyotard, *The Differend* and Michael Marder, *The Event of the Thing (Derrida's Post-Deconstructive Realism)*, (Toronto: University of Toronto Press, 2009).

2 In retrospect, the role of history in the event is a hangover from its dominant presence in *Theory of the Subject* where he says 'The double seal is the price of History for all novelty' (TS 125). Novelty is replaced by event in *Being and Event* whilst history falls away in favour of worlds in *Logic of Worlds*.
3 Early origins of the evental site are to be found in the 'empty place' which the subject forces to produce a 'splace' or state of novelty, through the destruction of a previous 'splace' or existing consistent world: TS 264. See also NN 61–3 which has a useful example using the philosophical communicable animal the cat, which is extensively commented on in Pluth, *Alain Badiou* 41–5.
4 Brian Anthony Smith, 'The Limits of the Subject', in Ashton, Bartlett and Clemens (eds), *The Praxis of Alain Badiou* 84–6
5 Ibid. 87.
6 Ibid. 88.
7 Hallward, BST 117–18.
8 Alain Badiou, *Logics of Worlds*, trans. Alberto Toscano (London: Continuum, 2009), 355–80. Hereafter referred to LW.
9 See also Gillespie, *The Mathematics of Novelty* 59–60 on this point.
10 Ibid. 65–6.
11 For an excellent explanation of self-predication see Alain Badiou *Second Manifesto for Philosophy,* trans. Louise Burchill (Cambridge: Polity, 2011), 78–80.
12 Again, see Gillespie, *The Mathematics of Novelty* on this point: 59–61, 65–6 and 68.
13 Yet it already exists as a central part of the materialism of *Theory of the Subject,* TS 218.
14 This is the basic summary of what *Logics of Worlds* concerns itself with.
15 The retroactive naming of the event is already a central element of Badiou's philosophy by the time of *Theory of the Subject*, see for example TS 126. And although Badiou moves away from nomination in *Logics of Worlds* he is still speaking of naming events in *St. Paul: The Foundation of Universalism*, trans. Ray Brassier (Stanford: Stanford University Press, 2003. Hereafter referred to as SP), for example, published a decade after *Being and Event* (SP 110). However, in an interview in *Infinite Thought* he is clear: 'It is not very good terminology, the terminology of the nomination. I now think that the event has consequences, objective consequences and logical consequences' (IT 129–30). This interview took place in 1999 just two years after the clear support for naming in *St. Paul* identifying the years 1997–9 as an important shift in Badiou's published thoughts on this issue. In contrast, it is clear in a book like *Wittgenstein's Antiphilosophy* that naming is a central component of the constructivism and anti-philosophy that he wants to replace with his logics of worlds, see WA 99–100, also Alain Badiou, *Manifesto for Philosophy*, trans. Norman Madarasz (Albany NY: SUNY Press, 1999). Hereafter referred to as MP, 94–5.
16 Ludwig Wittgenstein, *Philosophical Investigations* (Oxford: Blackwell, 1963), 89–104.

6 The Event, Intervention and Fidelity

1 See DCB 19–30. See TC 160–4 for a different angle in univocity.
2 The initial positing of the torsion of the event in *Theory of the Subject* answered precisely this worry. See TS 152.
3 See Alain Badiou, *Conditions*, trans. Steven Corcoran (London: Continuum, 2008. Hereafter referred to as C) 104 for a different approach to this issue and HI 129 and the discussion of the temptation of infidelity.
4 These qualities pertain to the possibility of not naming the event correctly so as to negate the event. For more on this see LW 369–78 and SP 12.
5 See Giorgio Agamben, *The Signature of All Things: On Method,* trans. Luca D'Isanto and Kevin Attell (New York: Zone Books, 2009), 9–32.
6 Habermas' communicative action simply takes Kant's communicability and makes it the foundation of subjectivity rather than a confirmation of a universal, objective subjective quality of judgement of taste. See Jürgen Habermas, *On the Pragmatics of Communication,* ed. Maeve Cooke (Cambridge: Polity, 1988).
7 As I mentioned, Habermas calls this communicative action but this lacks the radical, destabilizing element central to Badiou. A much closer concept is that of *klesis* in Giorgio Agamben, *The Time That Remains,* trans. Patricia Dailey (Stanford CA: Stanford University Press, 2005), 29–34.
8 For more on intervention see Bosteels 165–8 and Gibson's fascinating consideration of the relation between intervention and apogagic reasoning in Gibson, *Beckett* 63.
9 See also his comments in *Infinite Thought* 29–31 and Norris, *Badiou's Being and Event* 165–74 on Badiou and Pascal.
10 For the relation of Badiou to Pascal, in particular his wager, see Balibar, 'The History of Truth' in Hallward *Truth* 28, Barker, *Alain Badiou* 79–8 and Norris, *Badiou's Being and Event* 165–74.
11 See TS 286 for the origins of undecidability in Badiou and its relation to Gödel's theorem that some undecidables can be demonstrated. See also IT 46, TW 106–8 and MP 3–5.
12 There is work to be done to relate and differentiate Badiou's decisionism from that associated with the work of Carl Schmitt.
13 As I make clear in Watkin *Agamben and Indifference* 117–21.
14 See Laclau, 'An Ethics of Militant Engagement' in Hallward, *Think* 132 and 135 for a critical reading of Badiou's reliance on the yes/no construction.
15 See TS 314 for a discussion of deixis in Hegel's comments on the term 'this'.
16 In fact, in *The Handbook of Inaesthetics*, Badiou calls truth due to an event the 'unnameable' (HI 24). However, in the chapter on forcing in *Theoretical Writings* he seems to contradict this by calling the unforceable the unnameable, perhaps due to the negative influence of Lacan (TW 134). An interesting final consideration is Badiou's reading of the unnameable in relation to Beckett's great work of the same

name in Alain Badiou, *On Beckett*, Nina Power and Alberto Toscano (eds), (Manchester: Clinamen Press, 2003), 23–4. Hereafter referred to as OB.

17 See Brough's discussion of the famous distinction in Edmund Husserl, *On the Phenomenology of the Consciousness of Internal Time*, trans. John Barnett Brough (Norwell MA: Kluwer Academic Publishers, 2008), xl.

18 There are surprising parallels between Habermas and Badiou, over communicability but also the demand of freedom that determines such actions. See in particular Badiou's insistence that we must be free to choose in HI 53. For more on the ethics of forcing see TW 118 and 130.

19 For the origins of this conception see the chapter entitled 'Subjectivizing anticipation, retroaction of the subjective process' in TS 248–53.

20 A different, historical and political consideration of the Two can be found in TC 37.

21 I am deeply indebted to Andrew Gibson's work on this concept in his ground-breaking works *Beckett and Badiou: The Pathos of Intermittency* and *Intermittency: The Concept of Historical Reason in Recent French Philosophy* (Edinburgh: Edinburgh University Press, 2012).

22 In TS it is primarily the subject that is intervallic due to Lacanian algebra but the logic is very similar: 'The subject follows throughout the fate of the vanishing term, having the status of an interval between two signifiers ...' (TS 134).

23 Hallward, 'Introduction' in *Think* 13–17.

24 Jacques Derrida, *Acts of Literature*, ed. Derek Attridge (London: Routledge, 1992), 68.

25 The centrality of choice is there from the earlier work by Badiou, for example see CM 46 and 51.

26 See Norris, *Badiou's Being and Event* 138–43 and 174–83.

27 See Hallward, BST 339.

28 See Fraser 'The Law of the Subject' in Ashton, Bartlett and Clemens (eds), *The Praxis of Alain Badiou* 57 for more on Badiou's extensional approach.

29 Universality is not as pronounced a quality of fidelity here as it is across the entirety of *St. Paul: The Foundation of Universalism* where the concept of a universal singularity is often discussed. See in particular SP 11–15.

30 Alain Badiou, *Mathematics of the Transcendental*, trans. A.J. Bartlett and Alex Ling (London: Bloomsbury, 2014). Hereafter referred to as MT.

31 He already says as much in TS 274.

7 The Generic

1 See TS 267.

2 Hallward, BST 68–9. See also Bowden 'Set-Theoretical' in Bowden and Duffy, *Badiou and Philosophy* 52, Duffy 'Badiou's Platonism' in Bowden and Duffy, *Badiou and Philosophy* 68–70, and Barker, *Alain Badiou* 87 and 153.

3 Hallward, BST 69.
4 Ibid.
5 See Fraser, 'The Law of the Subject' in Ashton, Bartlett and Clemens (eds), *The Praxis of Alain Badiou* 50–1 for more on this discussion.
6 As Badiou says in *Theory of the Subject* 'The real is the impasse of formalization' (TS 22).
7 See Desanti, 'Some Remarks' in Hallward, *Think* for a consideration of the term generic as regards its relation to genesis and engendering, 65. For yet another sense of the generic as the reduction of complexity see OB 3.
8 It is in *St. Paul* that Badiou speaks of the 'genericity of the true' (SP 98).
9 In *Theory of the Subject*, Badiou speaks of two processes of knowledge in Marx, of its conditions and its seizing (TS 199). Although the Marxist inflections are later dropped, it is important to note that Marx is a primary source for the division truth/veridical in relation to knowledge as encyclopaedia. In contrast, in later work Badiou considers an opposition between fidelity and knowledge, see for example SP 45. See Gibson, *Beckett* 70 for more on this topic.
10 See MP.
11 To redress the balance as regards our neglect of the topic of conditions see Justin Clemens, 'Had we but worlds enough, and time, this absolute, philosopher . . .' in Ashton, Bartlett and Clemens (eds), *The Praxis of Alain Badiou* 111–20.
12 Badiou sees philosophy as sutured to four conditions: science, politics, love and art. It alone does not produce truths but, 'gathers them after the facts' (Oliver Feltham, 'Philosophy' in Bartlett and Clemens, *Alain Badiou: Key Concepts* 20) and does so only if it can show the compossibility of all four generic procedures. Indeed, if philosophy becomes too wedded to one condition as was the case with philosophy and politics in the twentieth century, the results can be disastrous. Truths occur across these different generic conditions regardless of philosophy's intervention yet to be identified as a truth qua truth, only philosophy has the vocabulary to do so. What this means is, as Hallward says 'The truths invented in love, art, science and politics are the conditions rather than the objects of philosophy' (BST 243). While we do not refute Badiou on this point, we are unconvinced of the stipulation of only four conditions, sometimes Badiou himself admits there could be more (Alain Badiou, *Second Manifesto for Philosophy*, trans. Louise Burchill (Cambridge: Polity, 2011), 21–2. Hereafter referred to as SMP. And more interested in other aspects of conditions such as their formal genericity. For more see MP 33–40 and Part 1 of C 1–132 and SMP 66.
13 For more on this see Duffy, 'Badiou's Platonism', in Bowden and Duffy *Badiou and Philosophy* 71
14 See IT 47.
15 See also Duffy, 'Badiou's Platonism' in Bowden and Duffy *Badiou and Philosophy* 73–4.
16 A useful summary of generic subsets can be found in IT 48–9.

8 Forcing: Truth and Subject

1 E 27.
2 Indeed, the definition of subject in the early work appears almost unrecognisable in relation to *Being and Event*. In *Theory of the Subject* Badiou says 'we will call *subjective* those processes relative to the qualitative concentration of force' (TS 41). Force, the central concept here, is hardly mentioned in the later work.
3 I was pleased to read Bosteel's analysis of the relation of *Theory of the Subject* to the two volumes of the Being and Event project. He calls *Theory of the Subject* 'a kind of vanishing mediator between the two volumes of *Being and Event*' (Bosteels 199), a position not dissimilar to my own and which I will investigate in more detail in the second volume of our study.
4 Chance, for example, plays a much more pronounced role in *Theory of Subject*, where it is one of a triplet of terms that determines the subject: chance, cause and consistency (TS 127). Although this triplet is not precisely the same as the structure of intervention in *Being and Event* one can see a very close parity between the two texts at this point.
5 See SP 84–5 for more on truth procedures.
6 Note its origins in TS 272–3. The literature on the concept of forcing is truly excellent. See Feltham, *Badiou* 108–10; Desanti, 'Some Remarks' in Hallward, *Think* 63 and 65; Duffy, 'Badiou's Platonism' in Bowden and Duffy, *Badiou and Philosophy* 70; Fraser, 'The Law of the Subject' in Ashton, Barlett and Clemens (eds), *The Praxis of Alain Badiou* 24; Anthony Smith, 'The Limits of the Subject' in Ashton, Barlett and Clemens (eds), *The Praxis of Alain Badiou* 91; and Bosteels, *Badiou and Politics* 174–90. Special mention should be made of Gillespie's excellent exposition on the topic, 83–5 and of course Hallward's essential introduction, *Badiou* 135–9.
7 In particular, see Gillespie, *The Mathematics of Novelty* 84–5 for a more detailed explanation of Cohen's concept of forcing in relation to the yes/no.
8 Although it has not totally disappeared, see for example C 54–6 and 131–2 where he recuperates destruction by clarifying its relational different to the now preferred term subtraction. See also the centrality of destruction in terms of the passion for the real that Badiou believes typifies the modern in TC 54–6. And the return to the concept in LW 394–6.
9 Badiou's early theory of the subject is expressed in the formula: 'Destruction is that figure of the subject's grounding in which loss not only turns lack into a cause, but also produces consistency out of excess' (TS 140). Destruction becomes the radical change of a world due to the subject's intervention and militancy as regards a truth, but change destruction to an event and the idea of the subject producing consistency out of excess is not far from the idea of the event in *Being and Event*. He clarifies the term's differing use in *Theory of the Subject* and *Being and Event* in IT 132.

10 Laruelle, *Anti-Badiou* xii.

11 Badiou is consistent in his attacks on Foucault's constructivism, see for example TS 188.

12 See LW 1–2.

13 Decisionism has always been a significant part of Badiou's work, for example in *Theory of the Subject* he says: 'The subject awakens to the decision, which is purely its mode of existence' (TS 172). We must be clear that the decision on the event pertains to the naming of the event not the event as such.

14 Watkin 12–15.

Bibliography

Agamben, Giorgio. 1991. *Language and Death: The Place of Negativity*, trans. Karen E. Pinkus with Michael Hardt. Minneapolis: University of Minnesota Press.

Agamben, Giorgio. 1999. *Potentialities,* trans. Daniel Heller-Roazen. Stanford CA: Stanford University Press.

Agamben, Giorgio. 2009. *The Signature of All Things: On Method*, trans. Luca D'Isanto and Kevin Attell. New York: Zone Books.

Agamben, Giorgio. 2005. *The Time That Remains,* trans. Patricia Dailey. Stanford CA: Stanford University Press.

Ashton, Paul, A.J. Bartlett and Justin Clemens (eds.). 2006. *The Praxis of Alain Badiou.* Melbourne: re:press.

Badiou, Alain. 1999. *Manifesto for Philosophy*, trans. Norman Madarasz. Albany NY: SUNY Press.

Badiou, Alain. 2000. *Deleuze: The Clamor of Being*, trans. Louise Burchill. Minneapolis: University of Minnesota Press.

Badiou, Alain. 2001. *Ethics: An Essay on the Understanding of Evil*, trans, Peter Hallward. London: Verso.

Badiou, Alain. 2003. *On Beckett*, eds. Nina Power and Alberto Toscano. Manchester: Clinamen Press.

Badiou, Alain. 2003. *St. Paul: The Foundation of Universalism*, trans. Ray Brassier. Stanford CA: Stanford University Press.

Badiou, Alain. 2004. *Theoretical Writings,* ed. and trans. Ray Brassier and Alberto Toscano. London: Continuum.

Badiou, Alain. 2005. *Being and Event*, trans. Oliver Feltham. London: Continuum.

Badiou, Alain. 2005. *Handbook of Inaesthetics,* trans. Alberto Toscano. Stanford CA: Stanford University Press.

Badiou, Alain. 2005. *Metapolitics,* trans. Jason Barker. London: Verso.

Badiou, Alain. 2006. *Briefings on Existence: A Short Treatise on Transitory Ontology,* trans. Norman Madarasz. Albany NY: SUNY Press.

Badiou, Alain. 2006. *Infinite Thought,* trans. Oliver Feltham and Justin Clemens. London: Continuum.

Badiou, Alain. 2007. *The Century,* trans. Alberto Toscano. Cambridge: Polity.

Badiou, Alain. 2007. *The Concept of the Model,* trans. Zachary Luke Fraser and Tzuchien Tho. Melbourne: re:press.

Badiou, Alain. 2008. *Conditions*, trans. Steven Corcoran. London: Continuum.

Badiou, Alain. 2008. *Number and Numbers*, trans. Robin Mackay. Cambridge: Polity.

Badiou, Alain. 2009. *Logics of Worlds*, trans. Alberto Toscano. London: Continuum.

Badiou, Alain. 2009. *Theory of the Subject*, trans. Bruno Bosteels. London: Continuum.

Badiou, Alain. 2011. *Second Manifesto for Philosophys*, trans. Louise Burchill. Cambridge: Polity.

Badiou, Alain. 2011. *Wittgenstein's Antiphilosophy*, trans. Bruno Bosteels. London: Verso.

Badiou, Alain. 2014. *Mathematics of the Transcendental*, trans. A.J. Bartlett and Alex Ling. London: Bloomsbury.

Badiou, Alain and Bruno Bosteels. 2005. 'Can Change be Thought? A Dialogue with Alain Badiou', in Gabriel Riera, *Alain Badiou: Philosophy and its Conditions*. Albany NY: SUNY Press, 237–62.

Barker, Jason. 2002. *Alain Badiou: A Critical Introduction*. London: Pluto Press.

Bartlett, A.J. 2006. 'Conditional Notes on a New Republic' in Paul Ashton, A.J. Bartlett and Justin Clemens (eds), *The Praxis of Alain Badiou*. Melbourne: re:press, 210–46.

Bartlett, A.J. 2010. 'Plato' in A.J. Bartlett and Justin Clemens (eds), *Alain Badiou: Key Concepts*. Durham: Acumen Press, 107–17.

Bartlett, A.J. and Justin Clemens. 2010. *Alain Badiou: Key Concepts*. Durham: Acumen Press.

Bartlett, A.J. and Justin Clemens (eds). 2010. 'Lacan' in A.J. Bartlett and Justin Clemens (eds), *Alain Badiou: Key Concepts*. Durham: Acumen Press, 2010, 155–67.

Bartlett, A.J., Justin Clemens and Jon Roffe. 2014. *Lacan Deleuze Badiou*. Edinburgh: Edinburgh University Press.

Balibar, Etienne. 2004. 'The History of Truth: Alain Badiou in French Philosophy', in Peter Hallward (ed.), *Think Again: Alain Badiou and the Future of Philosophy*. London: Continuum, 21–38.

Beaney, Michael (ed.). 1997. *The Frege Reader*. Oxford: Blackwell.

Bensaïd, Daniel. 2004. 'Alain Badiou and the Miracle of the Event', in Peter Hallward (ed.), *Think Again: Alain Badiou and the Future of Philosophy*. London: Continuum, 94–105.

Bosteels, Bruno. 2011. *Badiou and Politics*. Durham: Duke University Press.

Sean Bowden. 2012. 'The Set-Theoretical Nature of Badiou's Ontology and Lautman's Dialectic of Problematic Ideas' in Sean Bowden and Simon Duffy, *Badiou and Philosophy*. Edinburgh: Edinburgh University Press, 39–58.

Bowden, Sean and Simon Duffy. 2012. *Badiou and Philosophy*. Edinburgh: Edinburgh University Press.

Brassier, Ray. 2004. 'Nihil Unbound: Remarks on Subtractive Ontology and Thinking Capitalism', in Peter Hallward (ed.), *Think Again: Alain Badiou and the Future of Philosophy*. London: Continuum, 50–8.

Chiesa, Lorenzo. 2006. 'Count-as-one, Forming-into-one, Unary Trait, S_1' in Paul Ashton, A.J. Bartlett and Justin Clemens (eds), *The Praxis of Alain Badiou*. Melbourne: re:press, 147–76.

Clemens, Justin. 2006. 'Had we but worlds enough, and time, this absolute, philosopher . . .' in Paul Ashton, A.J. Bartlett and Justin Clemens (eds), *The Praxis of Alain Badiou*. Melbourne: re:press, 111–20.

Joan Copjec. 2005. '*Gai Savoir Sera:* The Science of Love and the Insolence of Chance' in Gabriel Riera, *Alain Badiou: Philosophy and its Conditions*. Albany NY: SUNY Press, 119–36.

Deleuze, Gilles. 1994. *Difference and Repetition*, trans. Paul Patton. London: The Athlone Press.

de Beistegui, Miguel. 2005. 'The Ontological Dispute: Badiou, Heidegger, and Deleuze', in Gabriel Riera, *Alain Badiou: Philosophy and its Conditions*. Albany NY: SUNY Press, 45–58.

Derrida, Jacques. 1978. *Writing and Difference*, trans. Alan Bass. London: Routledge.

Derrida, Jacques. 1982. *Margins of Philosophy*, trans. Alan Bass. London: Harvester Wheatsheaf.

Derrida, Jacques. 1992. *Acts of Literature*, ed. Derek Attridge. London: Routledge.

Desanti, Jean-Toussaint. 2004. 'Some Remarks on the Intrinsic Ontology of Alain Badiou', in Peter Hallward (ed.), *Think Again: Alain Badiou and the Future of Philosophy*. London: Continuum, 59–66.

Duffy, Simon. 2012. 'Badiou's Platonism: The Mathematical Ideas of Post-Cantorian Set Theory' in Sean Bowden and Simon Duffy, *Badiou and Philosophy*. Edinburgh: Edinburgh University Press, 59–78.

During, Elie. 2010. 'Art' in A.J. Bartlett and Justin Clemens (eds), *Alain Badiou: Key Concepts*. Durham: Acumen Press, 82–93.

Feltham, Oliver. 2008. *Alain Badiou: Live Theory*. London: Continuum.

Feltham, Oliver. 2010. 'Philosophy' in A.J. Bartlett and Justin Clemens (eds), *Alain Badiou: Key Concepts*. Durham: Acumen Press, 13–26.

Foucault, Michel. 1972. *The Archaeology of Knowledge*, trans. A.M. Sheridan Smith. London: Routledge.

Fraser, Zachary Luke. 2006. 'The Law of the Subject: Alain Badiou, Luitzen Brouwer and the Kripkean Analyses of Forcing and Heyting Calculus' in Paul Ashton, A.J. Bartlett and Justin Clemens (eds), *The Praxis of Alain Badiou*. Melbourne: re:press, 23–70.

Gibson, Andrew. 2006. *Beckett and Badiou: The Pathos of Intermittency*. Oxford: Oxford University Press.

Gibson, Andrew. 2012. *Intermittency: The Concept of Historical Reason in Recent French Philosophy*. Edinburgh: Edinburgh University Press.

Gillespie, Sam. 2008. *The Mathematics of Novelty: Badiou's Minimalist Metaphysics*. Melbourne: re:press.

Habermas, Jürgen. 1988. *On the Pragmatics of Communication*, ed. Maeve Cooke. Cambridge: Polity.

Hallward, Peter. 2003. *Badiou: A Subject to Truth*. Minneapolis: University of Minnesota Press.

Hallward, Peter. 2004. *Think Again: Alain Badiou and the Future of Philosophy*. London: Continuum.

Hallward, Peter. 2004. 'Introduction: Consequences of Abstraction', in Peter Hallward, *Think Again: Alain Badiou and the Future of Philosophy*. London: Continuum, 1–20.

Hegel, G.W.F. 1969. *Science of Logic*, trans. A.V. Miller. Amherst NY: Humanity Books.

Hegel, G.W.F. 1977. *Phenomenology of Spirit*, trans. A.V. Miller. Oxford: Oxford University Press.

Heidegger, Martin. 1969. *Identity and Difference*, trans. Joan Stambaugh. Chicago: University of Chicago Press.

Husserl, Edmund. 2008. *On the Phenomenology of the Consciousness of Internal Time*, trans. John Barnett Brough. Norwell MA: Kluwer Academic Publishers.

Kant, Immanuel. 1952. *The Critique of Judgement*, trans. James Creed Meredith. Oxford: Clarendon Press.

Kant, Immanuel. 2007. *Critique of Pure Reason*, trans. Marcus Weigelt. London: Penguin Books.

Laclau, Ernesto. 2004. 'An Ethics of Militant Engagement' in Peter Hallward, *Think Again: Alain Badiou and the Future of Philosophy*. London: Continuum, 120–37.

Laruelle, François. 2010. *Philosophies of Difference: A Critical Introduction to Non-philosophy*, trans. Rocco Gangle. London: Continuum.

Laruelle, François. 2013. *The Anti-Badiou: On the Introduction of Maoism into Philosophy*, trans. Robin Mackay. London: Bloomsbury.

Lecercle, Jean-Jacques. 2010. *Badiou and Deleuze Read Literature*. Edinburgh: Edinburgh University Press.

Ling, Alex. 2010. 'Ontology' in A.J. Bartlett and Justin Clemens (eds), *Alain Badiou: Key Concepts*. Durham: Acumen Press, 48–60.

Lyotard, Jean-Francois. 1988. *The Differend*, trans. Georges Van Den Abbeele. Minneapolis: University of Minnesota Press.

Machery, Pierre. 2005. 'The Mallarmé of Alain Badiou', in Gabriel Riera, *Alain Badiou: Philosophy and its Conditions*. Albany NY: SUNY Press, 109–18.

Marder, Michael. 2009. *The Event of the Thing (Derrida's Post-Deconstructive Realism)*. Toronto: University of Toronto Press.

May, Todd. 2004. 'Badiou and Deleuze on the One and the Many', in Peter Hallward, *Think Again: Alain Badiou and the Future of Philosophy*. London: Continuum, 67–76.

MacCannell, Juliet Flower. 2005. 'Alain Badiou: Philosophical Outlaw' in Gabriel Riera, *Alain Badiou: Philosophy and its Conditions*. Albany NY: SUNY Press, 137–84.

Mullarkey, John. 2010. 'Deleuze' in A.J. Bartlett and Justin Clemens (eds), *Alain Badiou: Key Concepts*. Durham: Acumen Press, 168–75.

Nancy, Jean-Luc. 2004. 'Philosophy Without Conditions' in Peter Hallward, *Think Again: Alain Badiou and the Future of Philosophy*. London: Continuum', 39–49.

Norris, Christopher. 2009. *Badiou's Being and Event*. London: Continuum.

Papineau, David. 2012. *Philosophical Devices: Proofs, Probabilities, and Sets*. Oxford: Oxford University Press.

Pluth, Ed, 2010. *Alain Badiou*. London: Polity.

Potter, Michael. 2004. *Set Theory and Its Philosophy*. Oxford: Oxford University Press.

Power, Nina. 2006. 'Towards an Anthropology of Infinitude: Badiou and the Political Subject' in Paul Ashton, A.J. Bartlett and Justin Clemens (eds), *The Praxis of Alain Badiou*. Melbourne: re:press, 309–38.

Power, Nina. 2012. 'Towards a New Political Subject? Badiou Between Marx and Althusser' in Sean Bowden and Simon Duffy, *Badiou and Philosophy*. Edinburgh: Edinburgh University Press, 157–76.

Power, Nina and Alberto Toscano. 2010. 'Politics' in A.J. Bartlett and Justin Clemens (eds), *Alain Badiou: Key Concepts*. Durham: Acumen Press, 94–106.

Rabaté, Jean Michel. 2005. 'Unbreakable B's: From Beckett and Badiou to the Bitter End of Affirmative Ethics', in Gabriel Riera, *Alain Badiou: Philosophy and its Conditions*. Albany NY: SUNY Press, 87–108.

Rancière, Jacques 2004. 'Aesthetics, Inaesthetics, Anti-Aesthetics', in Peter Hallward, *Think Again: Alain Badiou and the Future of Philosophy*. London: Continuum, 218–31.

Riera, Gabriel. 2005. *Alain Badiou: Philosophy and its Conditions*. Albany NY: SUNY Press.

Riera, Gabriel. 2005. 'For an "Ethics of Mystery": Philosophy and the Poem', in Gabriel Riera, *Alain Badiou: Philosophy and its Conditions*. Albany NY: SUNY Press, 61–86.

Riera, Gabriel. 2005. 'Introduction: Alain Badiou: The Event of Thinking' in Gabriel Riera, *Alain Badiou: Philosophy and its Conditions*. Albany NY: SUNY Press, 1–20.

Roffe, Jon. 2012. *Badiou's Deleuze*. Durham: Acumen Press.

Roffe, Jon. 2010. 'Spinoza' in A.J. Bartlett and Justin Clemens (eds), *Alain Badiou: Key Concepts*. Durham: Acumen Press, 118–27.

Ruda, Frank. 2015. *For Badiou: Idealism without Idealism*. Evanston IL: Northwestern University Press.

Shapiro, Stewart. 2001. *Thinking About Mathematics: The Philosophy of Mathematics*. Oxford: Oxford University Press.

Shapiro, Stewart (ed.). 2005. *The Oxford Handbook of Philosophy of Mathematics and Logic*. Oxford: Oxford University Press.

Shapiro, Stewart. 2005. 'The Philosophy of Mathematics and its Logic: Introduction' in Stewart Shapiro (ed.), *The Oxford Handbook of Philosophy of Mathematics and Logic*. Oxford: Oxford University Press, 3–27.

Smith, Brian Anthony. 2006. 'The Limits of the Subject in Badiou's Being and Event', in Paul Ashton, A.J. Bartlett and Justin Clemens (eds.), *The Praxis of Alain Badiou*. Melbourne: re:press, 71–101.

Smith, Daniel W. 2004. 'Badiou and Deleuze on the Ontology of Mathematics' in Peter Hallward, *Think Again: Alain Badiou and the Future of Philosophy*. London: Continuum, 77–93.

Steinhart, Eric. 2009. *More Precisely: The Math You Need To Do Philosophy*. Peterborough Ont.: Broadview Press.

Tho, Tzuchen. 2012. 'What is Post-Cantorian Thought? Transfinitude and the Conditions of Philosophy', in Sean Bowden and Simon Duffy, *Badiou and Philosophy*. Edinburgh: Edinburgh University Press, 19–38.

Tiles, Mary. 1989. *The Philosophy of Set Theory: A Historical Introduction to Cantor's Paradise*. Mineola NY: Dover Publications.

Watkin, William. 2015. *Agamben and Indifference.* London: Rowman & Littlefield International.

Wittgenstein, Ludwig. 1963. *Philosophical Investigations.* Oxford: Blackwell.

Zizek, Slavoj. 2004. 'From Purification to Subtraction: Badiou and the Real' in Peter Hallward, *Think Again: Alain Badiou and the Future of Philosophy.* London: Continuum, 165–81.

Index